# Physics Over Easy
## Breakfasts with Beth and Physics

### 2nd Edition

# Physics Over Easy

## Breakfasts with Beth and Physics

**2nd Edition**

## Leonid V Azároff
University of Connecticut, USA

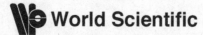
**World Scientific**

NEW JERSEY · LONDON · SINGAPORE · BEIJING · SHANGHAI · HONG KONG · TAIPEI · CHENNAI

*Published by*

World Scientific Publishing Co. Pte. Ltd.

5 Toh Tuck Link, Singapore 596224

*USA office:* 27 Warren Street, Suite 401-402, Hackensack, NJ 07601

*UK office:* 57 Shelton Street, Covent Garden, London WC2H 9HE

**British Library Cataloguing-in-Publication Data**
A catalogue record for this book is available from the British Library.

First published 2010
Reprinted 2011

**PHYSICS OVER EASY (2nd Edition)**
**Breakfasts with Beth and Physics**
Copyright © 2010 by World Scientific Publishing Co. Pte. Ltd.

ISBN-13 978-981-4295-44-4
ISBN-10 981-4295-44-2
ISBN-13 978-981-4295-45-1 (pbk)
ISBN-10 981-4295-45-0 (pbk)

Typeset by Stallion Press
Email: enquiries@stallionpress.com

Printed in Singapore.

*Dedicated to*
Diana, David, Richard, and Lenore

# Contents

# Contents

# Preface to the Second Edition

It sometimes occurs to readers of the original edition of *Physics Over Easy* to wonder whether the reported breakfast conversations actually took place or whether they were invented by the author. Let me reassure you, they did take place as originally reported in the Introduction. So when I was invited to write an updated second edition, I wondered how I could continue maintaining that format. Since both my wife Beth and I had retired, our lifestyle changed and so did the circumstances that prompted the original discussions. Nevertheless, when she learned that I had agreed to write a second edition, Beth insisted that I brief her on the new material as I covered it. Thus we continued to conduct periodic conversations on physics, but they were no longer limited to breakfasts but also included our other meals. Slightly edited versions of these discussions are reported in Chapters 19 through 24. The earlier parts of the book are pretty much unchanged from the original, other than for some minor corrections.

The new material describes more recent discoveries that matured during the past decades. Most are still ongoing and should be viewed as work in progress. The original book did not include them because our understanding was too incomplete at the time of its writing. This is an exciting time for physicists. New and more sensitive probes are being sent aloft on satellites to study the cosmos. More powerful accelerators are coming on line to look for the Higgs and solidify our understanding of the subatomic world. Experimentalists are seeking practical applications of nanoscale materials and probing ways to use the weirdness of quantum mechanics in useful devices and theorists are seeking ways to expand Grand Unified Theories (GUTS) to encompass all of physics. I hope that you'll enjoy perusing a glimpse of these exciting developments.

*Leonid V. Azaroff*

# Introduction:
# What Keeps Us Going?

## *Physicists, and Other People*

By now I have learned not to admit that I am a physicist. When a new acquaintance, sooner or later, asked me what I do, I used to answer, "I am a physicist." Then I would watch the eyes glaze over as my questioner looked for some graceful way to escape. So, after a long while, I decided not to admit being a physicist. Instead I would say that I am a materials scientist. This answer, somehow, proved to be either less frightening or more intriguing. Typically it would elicit a follow-up question of some sort. To "What does a materials scientist do?" I might reply that I was the director of a large research institute at the University of Connecticut, which studied the properties of all kinds of materials, and then I would quickly change the subject.

If asked what materials science *is*, I might explain that it is the study of the physical properties and underlying structure of the materials from which everything in the world is made, including people. At that point, usually, the questioner would change the subject.

If the new acquaintance pressed for more details, I would explain that my personal interest lay in discovering the physics of individual atoms and how they determined the properties that materials may exhibit. In return, I would be informed that my new friend admired physics or, perhaps, stood in awe of physicists. For a variety of reasons, however, he or she had not pursued the subject. If questioned still further, I sometimes admitted having authored several physics texts. This admission was

1

frequently followed by a wistful exclamation: If only I would write a book that they could understand!

What every physicist secretly dreads occurred to me a couple of summers ago. I was awaiting having my throat cut (well, neck, actually) at the Yale-New Haven Hospital. The surgeon, who was about to perform the carotid endarterectomy, asked me in the preoperation waiting room what I did. Upon learning that I was a physicist, he gazed down on me sympathetically and admitted that he had great difficulty getting into medical school because he nearly flunked his physics course in college. Now you understand why it is not a good idea to admit to strangers that one is a physicist!

Women tended to be my most sympathetic questioners. This was particularly true if they were single and discovered that I was also unmarried. While on a cruise in the Caribbean, I met a woman who was so sympathetic that I proposed marriage to her even before the cruise had finished.

Beth was an accomplished psychologist specializing in the behavior management of school children. A professor of physics was hardly a challenge to a successful tamer of bratty kids. Yet she never let me know this and always evinced the greatest respect for one who was learned in a subject that she had somehow neglected studying during her extended educational career.

Beth had completed her doctoral program at the University of Minnesota. Although she had managed to spend nine years in college without studying physics, her training in science included a philosophy of science course in which she discovered that the laws governing nature were parsimonious: the simplest possible explanation is usually the correct one for any natural phenomenon. Even without direct exposure to it, Beth realized that physics, being the most basic science, must provide parsimonious explanations. So she now harbored some regrets about not knowing any physics. Very curious by nature, she welcomed information about how an electric fuse worked, or how a rocket could fly, or why the sky was blue all day but turned red after sunset. She would be glad to linger at the breakfast table in order to unravel some such mysteries.

Whereas Beth, as a young woman, had been discouraged from pursuing subjects like physics, my own early educational experiences had been much more positive. Finding all school subjects, and especially mathematics, very easy, I relished the attention and encouragement lavished on me by my teachers. This pleasure was lessened only slightly by occasionally having to defend my right to being a teacher's pet with my far less talented fists. Referred to repeatedly as 'that little Einstein', I grew up with a burning desire to become like Einstein — a physicist. Of course, I had no idea what being a physicist meant other than it gave me a distinction among peers who aspired to become accountants, or doctors, or lawyers, or, occasionally, baseball players. By the time I went to college and majored in physics, I realized that the science of physics actually was neither more or less interesting or challenging than any of the other subjects pursued by my fellow students, but it held many fascinations for me. It wasn't until I had slogged through all required courses in graduate school that I fully appreciated why being a physicist was not synonymous with being an Einstein. But, by then the die had been cast.

Why does physics inspire such dread? The truth is that physics is no more difficult than any other college subject. What probably makes it *appear* to be difficult is the mystique that nonphysicists attribute to it and that physicists love to perpetuate. I also discovered, after becoming a physics teacher myself, how much easier it is to dwell on the intricacies of the mathematical language that physicists use than it is to explain the underlying meanings of all those equations to students. Unfortunately, it is often poorly prepared teachers who make any subject seem to be difficult or uninteresting.

My own *bete noir* in college was a brief and most unsatisfactory exposure to poetry. I have not really liked poetry ever since. Yet I have also heard skilled expositors who could make even poetry sound appealing!

So it was that some time after I married Beth, I decided to try my hand at writing a book that would make physics easier and more interesting for general college students. As I became increasingly absorbed in this effort, it spilled over into our breakfast conversations. In this way I became the physics Sheherezade of a thousand and one breakfasts.

What follows are my recollections of some of the more memorable of these breakfast discussions...

# Chapter One:
# Breakfast of Hard-Boiled Eggs
# with Inertia

## How It All Began — with Rolling Balls

Our Breakfast discussions of physics began one morning when I observed: "I am really enjoying my review and discovery of many tidbits of information about the early physicists."

This naturally prompted Beth to ask "Who *was* the first physicist?" (Her boundless curiosity prompts Beth to pose many such questions, as will be soon apparent.)

"Galileo Galilei was the first physicist, although he himself did not realize that. You see," I added, "scholars were not labeled 'scientists' or 'physicists' until the middle of the nineteenth century when the Master of Trinity and occasional Vice Chancellor of the University of Cambridge, William Whewell, first proposed such a name."

"But surely there were others who had studied the topics we now call physics long before Galileo," Beth protested. "How about the many advanced civilizations of the Egyptians, or Greeks, or Mayans, or Aztecs, or those in Asia and elsewhere?"

"We can find the remains of ancient structures in the Americas, in Asia, and in Africa, and we know about the many simple tools that the early builders had already discovered and utilized." I responded. "But, surprisingly, the earliest records we have of scientific analyses go back to the teachings by ancient scholars in what we now call Greece and Turkey. The sole exception was the systematic observation of heavenly bodies, but this was more an art than a science. Also, its end use was directed at

5

predicting the fortunes of rulers, not toward understanding the workings of nature.

What did take place back then, however, has direct bearing on Galileo's life. Beginning about 2,500 years ago, some of the people living along the shores of the cerulean seas surrounding what we now call the Near East began to speculate out loud about their ideas on what made up their environment. Maybe because they did not call themselves scientists, they had no trouble attracting others to listen to these ideas and to join in their efforts."

"Lee, are you suggesting that, if professors did not call themselves professors, they would get more students to listen?"

"It's an interesting possibility. In any case, 'schools' developed gradually under the tutelage of individual Greek scholars. One of the earliest, Pythagoras, now called the father of *natural philosophy,* is best remembered for deducing that the sum of the squares of the two short sides of a right triangle just equals the square of the long side, called the hypotenuse. Less well known is his claim that the earth must be spherical, at a time when everyone else was convinced that the earth was perfectly flat. Pythagoras did not base his conclusion on an observation that boats reaching the horizon gradually disappear from view (Fig. 1) but rather on the premise that a sphere was the most aesthetically pleasing shape a body could have. For this reason alone, a perfect earth should be a sphere! To protect themselves from popular anger at such an unorthodox idea, however, the Pythagoreans formed a secret society, swearing not to divulge their conclusions to outsiders."

"Were they more concerned about being ridiculed or about their personal safety?" Beth continued.

"I don't really know. It may be that forming a secret society was just a matter of intellectual snobbery." I responded before going on.

"You doubtlessly remember from your course in logic that the rules for rigorous reasoning were systematically spelled out by the most brilliant Greek philosopher, Aristotle. He then applied these rules to the analysis of a broad range of topics of possible interest to humans. In doing this, he never felt a need to subject his conclusions to any test other than that the arguments leading up to them obey these rules of logic. What puzzles me, however, is how the same people who required such rigorous proofs for

Fig. 1.   As a boat passes over the horizon (defined by the straight line of vision of an observer), the spherical shape of the earth causes a boat to 'sink' out of view, in the same way as the sun does at sunset.

their theorems in geometry could have been so cavalier in their acceptance of untested ideas about everything else?"

"I believe that Aristotle's writings included the description of human behavior." Beth noted. "That would make him the very first psychologist as well."

"Aristotle had fixed ideas about a lot of things, including how bodies on earth move. First he distinguished between what he called 'natural' motion and 'violent' motion. Natural motion was supposedly based on the 'nature' of an object. An object whose nature was of the earth, for example, a stone, would tend to return to the earth when lifted up and released. The larger or heavier the stone, the more rapidly would it seek to return. Similarly, smoke rises from a burning log because smoke is of air and so it seeks to return to the air. Such natural motion could take place in a straight line up or down, as was the case for most natural motion on earth."

"What about a leaf?" Beth asked. "It floats rather than falls to the earth in a straight line."

"According to Aristotle, a leaf's nature is mostly of earth but also partly of air. It should fall to the ground like a stone, but much more slowly."

"How did he handle the motion of the planets and the stars?" Beth wanted to know next. "They don't travel in straight lines toward or away from the earth."

"Here again Aristotle provided a neat answer. Celestial objects move in circles. Since circular motion has neither a beginning nor an end, it can go on forever."

"Like circular conversations?"

"That's probably why we call them circular. But let me continue. To move a stone in a horizontal direction, a 'violent' motion had to be imposed on it. Such violent motion could be imparted by a rapidly moving stream to a log floating on it, or by the wind to a falling leaf or to the sails of a ship. When the external agent imposing such violent motion is removed, the object returns to its natural state. The stone falls back to the earth and the log and boat float in place without further lateral motion."

"Wait a minute," Beth interjected. "When a stone is thrown sideways, the hand imparts a violent motion to it. But how did he deal with the fact that the stone keeps moving sideways after the hand releases it?"

"Here is just one example of Aristotle's genius: as the stone moves forward, he reasoned, it must push the air ahead of itself out of the way and leave a pocket devoid of air behind it. The compressed air, therefore, rushes toward the back of the stone to fill this void. In doing this, it propels the stone forward! Gradually, the stone strives to return to its natural state and falls back to the earth, from whence it came. How do you like that as a neatly logical explanation?"

"O.K. Can we agree that Galileo may be called the very first physicist but that others earlier laid down some of the groundwork?" Beth asked as she plopped two eggs into boiling water in a saucepan. "Now let's get back to where we started. Tell me more about Galileo and what he did to distinguish himself."

"Galileo was born into an impecuneous but noble family in Pisa, Italy on February 15, 1564. His father had a deep interest in the mathematical aspect of music but urged his most talented son to pursue a financially more rewarding career in medicine. So the young Galileo enrolled at the University of Pisa to prepare himself for a life as a physician. Like his father before him, however, Galileo found the study of mathematics far more interesting. He dropped out of the medical program to indulge his scientific curiosity and, fortunately, the rectors of the University of Pisa recognized his scientific promise and appointed him an instructor in mathematics. They did this even though Galileo held no formal degrees in mathematics or in any other field. This enabled him to stay on for three additional years while he did his seminal work on the motion of freely falling bodies there."

"What had Galileo done that so impressed his elders?" Beth wanted to know.

"Already while pursuing his medical studies, Galileo was frustrated by the inadequacies of Aristotle's teachings. One day while attending a church service in Pisa, Galileo was fascinated by the regularity of the back-and-forth swinging of the chandelier suspended from the ceiling by a long chain. Timing the duration of each swing against his own pulse beats, he found each swing to last exactly as long as the one before it. Later, at home, he duplicated this observation by suspending various weights from strings of different lengths. He found that he could change the duration of each swing by changing the length of the string or the weight of the suspended object. Aristotelean logic could not explain this. Neither could Galileo come up with a very satisfactory explanation. This did not prevent him from inventing a *pulsometer*, however, which is the forerunner of the metronomes that would annoy music students for many years to follow."

"Isn't a swinging candelabrum simply a pendulum?" Beth interjected. "And what distinguished Galileo from Aristotle? Was it the fact that he was not satisfied just speculating about the pendulum's regularity and instead carried out actual experiments?"

"That's part of the distinction and a most important part." I responded "But another, even more important distinction was Galileo's ability to generalize his experimental observations into a general law of nature. To see how that happened, let me describe for you how he corrected the misconception about freely falling objects that persists among some people even to the present day."

"You mean the one derived from Aristotle's conclusion that what goes up must also come down?" Beth asked mischievously as she plopped another pair of eggs into the saucepan.

"Not exactly. Aristotle taught that the speed of a falling object depended on its weight — the heavier the object, the faster it should fall toward the earth. To test this hypothesis, Galileo did not climb atop the leaning tower of Pisa, as some legends suggest. He realized that comparing the descent of two different objects would be fraught with innumerable difficulties: they have to be released at exactly the same time. The times of their release and their arrival below must be clocked precisely. Additionally, an allowance would need to be made for the resistance of the air through which they were falling. We are all familiar with the effect that streamlining has on the speeds of cars, boats, etc., and Galileo was not unaware of the role that air would play in free fall. He also lacked a stop watch or any time piece, for that matter, with which he could measure sufficiently accurately the speed of an object's descent."

"I hope that you make this clear in your textbook, Lee. It's hard for a reader in the twentieth century to realize that there was a time without watches or the many other things that we take so for granted."

"I try to do this by describing how Galileo overcame these limitations. First of all, he had a set of very smooth boards made up which contained V-shaped grooves so that he could roll a ball along the groove in a perfectly straight line. (The groove also minimized contact with the rolling ball, that is, the effect of friction.) By inclining a board at different angles, Galileo could vary the speed of a ball's descent in a controlled manner. To measure that speed, he arranged little tripping devices that rang a bell as the rolling ball passed by. Finally, he controlled the flow of water from a reservoir into a small container by putting his finger over the outlet tube of the reservoir. When the ball tripped the first bell, he released the water and, when the second bell rang, he shut it off again. Careful weighing of the water in the container then provided an accurate measure of the amount of time elapsed between the two consecutive events. To be sure that his measurements were reliable, Galileo repeated each trial a number of times. Most importantly, he kept accurate records of all his measurements."

"I know all about the importance of keeping accurate records," Beth exclaimed. "But what, exactly, did Galileo learn from watching balls roll down an inclined plane?"

"The first thing that Galileo discovered as he timed the descent of his freely rolling balls was that their speed did not depend on their weight, as Aristotle had incorrectly claimed. In fact, he found that the speed of all balls, regardless of their size or weight, increased in a very regular way: the further they rolled, the faster they rolled. This increase in speed is called *acceleration*, by the way.

As importantly, Galileo noted that the acceleration of every ball remained constant regardless of what the angle was at which he inclined the board! To check this out further, he measured the speeds during different intervals of a particular roll and again observed that the acceleration, that is, *the rate at which the speed increased*, was the same regardless of how fast the ball was rolling at the start of any measurement. The only thing that varied a ball's acceleration was the inclination angle of the board. The steeper the board, the faster the ball descended."

"I realize that this can't be actually done in practice" Beth interrupted, "but what would the acceleration be if the board were absolutely vertical?"

"That is a very pertinent question," I responded. "Should the board be raised to a vertical position, the constant acceleration measured by Galileo would become the same as the constant acceleration of a freely falling body!

In fact, this is what distinguishes Galileo from Aristotle. Instead of being satisfied with a logical explanation of his observations, Galileo asked, what would the result have been in an ideal experiment? How would the different size balls behave if they did not have to roll on actual boards? In this way, Galileo was able to extend his observations of rolling balls to the free fall of any object regardless of its size of shape, as long as it could be imagined falling through a vacuum that offered no air resistance."

"Did Galileo know about vacuums?" Beth inquired.

"The idea of a vacuum goes back to the Ancient Greeks, but how well Galileo understood what a vacuum is or whether one actually exists anywhere in the world, I really don't know." I replied. "Galileo did understand that a leaf or piece of paper falls to the ground with a different acceleration than a stone because a leaf encounters more air resistance than a stone."

"Well, are you suggesting that the shape of a freely falling object affects its rate of descent?" Beth persisted. "I thought that you said a moment ago that the acceleration of any object in free fall was the same regardless of its size or its shape,"

"As long as the object is not encountering air resistance, neither its shape nor its size, nor its weight make any difference." I rephrased my explanation.

"The reason we call Galileo the very first physicist is that he was the first person to generalize his observations in the actual world to an ideal world in which there were no impediments to the completely free fall of an object. It is this ability to extract the general rules that govern behaviors in an ideal world, before modifying them to fit the real one, that distinguishes physics from the other kinds of intellectual pursuits of humans."

"Then I don't understand why there is doubt that Galileo actually threw stones from the leaning tower in Pisa," Beth observed. "Surely a heavy stone and a smaller, lighter one would have encountered nearly the same air resistance."

"You're quite right. Both stones should reach the ground at essentially the same time." I responded. "Although there is no documental evidence, it has been suggested that Galileo did, in fact, perform such a demonstration because he was quite a publicist of his works. For example, Galileo published his technical results in Italian so that lay persons could read them. This, alone, made him unpopular with his colleagues who insisted that scholarly works should be published in Latin."

"But we're getting away from his discoveries about motion." I continued. "In a different experiment, Galileo arranged the ball rolling down one inclined board to continue rolling on a similarly grooved horizontal board and then to roll upward onto a third board inclined from the horizontal. If the two end boards were inclined at the same angle to the horizontal one, the ball would roll up on the third board to virtually the same height as the one from which it descended on the first board (Fig. 2). If the last board were less steeply inclined, then the ball would need to roll much farther to reach the same height. What would happen, Galileo wondered, if the ball rolled on a horizontal board that was endless and extended forever? From his previous measurements and careful notes, he deduced that the only

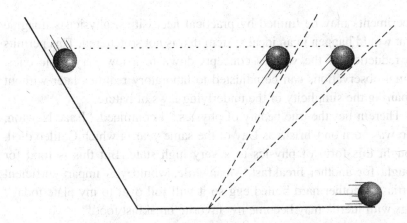

Fig. 2.   A ball rolling down an inclined board on the left will continue rolling up the inclined board on the right until it reaches the same height, regardless of its inclination. What would happen, Galileo wondered, if the board on the right remained horizontal for ever?

thing that ultimately caused a ball to stop rolling appeared to be a friction between the ball and the board. Without such a hindrance, Galileo generalized, *an object set in motion continues to be in motion while an object at rest continues to be at rest unless acted upon by an external agent.* This concept, called *inertia*, clearly contradicted the teachings of Aristotle and caused Galileo no end of grief in seventeenth-century Italy. While Aristotle saw no need for a frictionless board for a ball to roll on, Galileo's genius lay in his ability to generalize from his laboratory experiments to an *idealized* situation in which a rolling ball can go on rolling for ever."

"Are you saying that Galileo was more imaginative than Aristotle?" Beth interjected.

"No, because I believe that Aristotle had to be far more imaginative since he had no experimental facts on which to base his conclusions."

"Well," Beth persisted, "are you saying now that it takes more imagination to fantacize than to deduce the truth about nature?"

"Maybe I should have called Aristotle more speculative than Galileo." I replied. "Scientists and, especially, physicists have continued to exercise their imaginations to the present day. Although their laboratory experiments may be limited by practical necessities, physicists imagine what would happen in an ideal system that is not so fettered. This permits the reduction of the various concepts down to a few very basic ones. Actual observations can be adjusted to laboratory realities later without impairing the simplicity of the underlying laws of nature."

"Therein lies the true beauty of physics." I continued. "Isaac Newton, who was born on Christmas Day of the same year in which Galileo died, brought this form of physics to a very high state. But this is food for thought for another breakfast. Meanwhile, would you impart sufficient inertia to another hard boiled egg so it will roll over to my plate today? Eggs with inertia may become my favorite breakfast food!"

# Chapter Two:
# Breakfast of Eggs Bene-Bricked

## *What Keeps the Ball Rolling?*

"How would you like your eggs this morning? With inertia?" Beth inquired.

"Could I have some eggs Benedict, please?" I replied.

"You promised to tell me more about motion." Beth said some time later, as I forked a forkful of eggs in golden sauce.

"Sure" I responded. "Galileo did much more than study the acceleration of freely falling bodies to earn his title as the father of physics. For example, he built one of the very first telescopes that enabled him to discover the moons of Jupiter and make many other significant observations. Galileo also built one of the earliest thermometers for measuring air temperatures, and so forth. But for our purposes today, his discovery of inertia and the impact of this discovery is what matter most."

"Did Galileo ever wonder why an object had inertia or why it always fell to the earth with a constant acceleration?" Beth wanted to know.

"If he did, his musings have not been recorded. It was left to Isaac Newton to ponder the cause of such motion. Newton did just that while retreating to his parental home to escape the Great Plague that had closed the University of Cambridge shortly after he received his Bachelor of Arts degree there. Actually Newton had at least two previously conceived ideas about motion to consider. In addition to Galileo's, there were some novel proposals in the *Principia Philosophiae* that the great French philosopher and mathematician, Rene Descartes, had published when Newton was only two years old. In Part II of this treatise, Descartes talks about a *quantity of motion.* Moreover, he insisted that this quantity must be conserved, that is, remain constant throughout the universe!"

"At a time when most people still thought that the earth was flat and lay at the center of the entire universe, that was quite an assertion to make." Beth noted. "How did Descartes justify it?"

"Using his own words, excerpted from an English-language translation of the *Principia*, Descartes attributed the cause of motion to 'nothing other than God, Who in His Almighty power created matter with motion ... and Who thereafter *conserves in the universe* by His ordinary operations *as much of motion and rest as He put it in the first creation*' (the Italics indicate the special emphasis I placed on those words). Not only did God create all motion in the universe, but having created it, He keeps the total amount of motion and rest constant! This Descartes argued must be so because we must not 'attribute to Him inconstancy.'"

"Sounds like Descartes, not unlike Aristotle, preferred a well-reasoned argument to the experimental approach of Galileo." Beth observed.

"That is quite true. Nevertheless," I continued, "the idea that the total quantity of motion must remain constant presented a radically new way of thinking and one that set a pattern for much of the physics to follow. First of all, to demonstrate how the quantities of motion could be conserved, it is necessary to consider two or more bodies moving simultaneously. Until then, Galileo and others were content with analyzing the motion of a single object alone. Secondly, the idea that any quantity in nature should remain unchanged throughout the universe was a concept not heard before. Today such conservation laws lie at the very heart of physics. We cannot prove these laws by logical arguments nor derive them from any laws of nature that are more fundamental. So we accept conservation laws as self evident truths, have an abiding faith in them, and continuously use them to explain our various observations of nature."

"Let's get back to Newton," Beth interposed. "you said that he had learned from Galileo that the inertia of a body controls its state of motion. From Descartes he received the idea that the total quantity of motion in the universe must remain constant. But what exactly is this quantity of motion?

"We now call this quantity of motion the *momentum* of a moving body. When a freely falling body increases its speed at a constant acceleration, as Galileo had demonstrated, then its momentum is continuously increasing, as Descartes might have phrased it."

"But how does this fit the law of inertia which states that an object set in motion must continue in that same motion unless acted upon by some external force?" Beth asked.

"This was left to Newton to answer. It was his ability to connect all such ideas that set Newton apart from his contemporaries. His exact thought processes are, of course, unknown to us. It is reasonable to assume, however, that Newton began by accepting Galileo's law of inertia as one of the fundamental laws of motion. At least he listed it first in his subsequent publication. Next he argued that in order to produce a change in this steady state of motion or rest, an external agent or *force* must act to speed it up or to slow it down. We call this change in speed the acceleration of the object. Consequently, acceleration is related to the change in the quantity of motion of an object. We say nowadays that force equals *the rate of change of momentum.* But we know that acceleration is also the rate of change of speed. Thus it must follow that one of the factors contributing to momentum is the speed of the moving object. Actually this had been recognized already by Descartes who stated that the momentum of an object is equal to its weight multiplied by its speed."

"Six-year olds don't weigh much, but they sure can make up for it with their speed."

"Exactly!" I responded. "Incidentally, Descartes' definition of momentum troubled Newton. Any object, say an ordinary brick, has a certain weight. Although it may be resting on a table, should the table be yanked out from under it, the brick will fall to the floor. Suppose another identical brick is placed on top of the first one. Removing the table now will cause both bricks to fall to the floor together. But will the acceleration of the paired bricks be the same as of the single brick? According to Galileo's observations that rolling balls having different sizes reach the ground at the same time, the answer is: yes, a pair of bricks in free fall feel the same acceleration that a single brick does. This means that the weight of a brick is a measure of the external force that causes a brick to accelerate or change its momentum while it is falling to the ground. A double brick, weighing twice what a single brick weighs, feels a doubled force acting on it to produce the same acceleration. Some inherent property of a brick, therefore, must be the determining factor of what the force acting on it is or what its momentum should be. What this property of matter may be is not at all obvious and it puzzled Newton greatly. As soon as he figured out an acceptable answer, Newton incorporated it in his second law of motion."

"Six-year olds would enjoy seeing the brick experiment," Beth interrupted. "How are your eggs Benebricked, I mean, Benedict?"

"Just fine! Returning to the brick resting on the table," I continued, "what keeps it from pushing through the table top and falling to the floor? If the brick has associated with it a downward force called its weight, then the only way it can obey the law of inertia requiring it to remain at rest is for the table to exert an equal and opposite force on the brick! Newton quickly recognized that this meant that *for every action there must be an equal and opposite reaction* and listed this as his third and final law governing all motion."

"Hold on a minute!" Beth exclaimed. "I am getting confused. If two bricks weigh twice as much as one brick, why is not their weight an inherent property of the bricks?"

"Thanks to Newton's later works, we now realize that the weight of an object is simply a measure of the force that the earth's gravitational pull exerts on it. But this discovery was made by Newton somewhat later. Recall that the analysis of motion was carried out by him during a couple of years following his graduation from college. Without appreciating the cause of the force that produced weight, Newton realized even then that it had to be an external force that caused an object's acceleration or change in momentum to take place."

"I'm still not clear," said Beth. "If weight is an external force, what is the intrinsic property of an object on which this force called weight is acting?"

"Your question turns out to be as profound as asking: who is this God to whom Descartes attributes all motion?

Newton's conclusion regarding this intrinsic property of matter appears in the very first definition that he lists as a preamble to describing his three laws of motion. He calls this inherent property of matter its *mass,* but then he gives it a circular definition. Quoting from the first English translation of the original Latin text, Newton states that, for any object, 'the quantity of matter is the measure of the same, arising from its density and bulk (volume) conjunctly (multiplicatively).' The trouble with this definition is that what we call the density of an object is defined as its mass divided by its volume, so that, naturally, its mass must equal the density multiplied by its volume."

"Isn't that argument rather crafty?" Beth asked. "It seems somewhat circular to me."

"It certainly is. What Newton stressed later was that it may be better to think of this *mass* as an intrinsic property of matter that *does not change under the action of external forces.* He went on to point out that, if you double the density, you double the mass. If you double the volume, you double the mass. If you double the density *and* double the volume, the mass increases fourfold, and so forth. Obvious as this may be, it still doesn't tell us what mass actually is. It is probable that Netwon meant to suggest that matter is the embodiment of mass. He certainly recognized that *mass is the property of an object that causes it to resist any change in its state of motion.* Called the inertial mass of a body, however, it still doesn't tell us what mass is."

"So we just have to take it on faith?"

"Quite so. This is merely the first basic quantity in physics that we have to accept as given. We can't feel, or touch, or actually see the mass of an object in the same way that we can, feel, touch, or see the object itself. Yet the mass of an object, by virtue of its invariance, is an extremely useful concept that underlies most other basic considerations in physics. And don't forget that the laws of physics govern throughout our universe and underlie the behavior of all the matter it contains."

Noting a doubtful look on Beth's face, I continued: "Let me try to make the concept of mass clearer. The mass of any body, including you and me, does not change when acted upon by an external force. That's why we say that it is an intrinsic property of all matter. The external force of gravity, which we call weight, acts on this mass by imparting to it a constant acceleration during free fall. This force, or weight, must equal, therefore, the mass of an object times this constant gravitational acceleration. This is actually one way of stating Newton's second law of motion. As we know from having watched our astronauts cavorting on the moon. Because the moon's gravitational force is considerably less than that of the earth, their weight on the moon is correspondingly less also. Their mass has not been changed by this change in gravity, because it is an intrinsic property of matter. Removed from the gravitational attraction of any large body, an astronaut flying in outer space is virtually weightless but the astronaut's mass remains fully intact."

"Well I accept the concept of mass," Beth admitted, "but I still don't see why I can't detect it more directly than by having faith in the laws of physics."

"I hate to give undue credit to poets, but I believe it was some poet or other who coined the phrase 'sweet mystery of life.' Let's just accept that mass is part of the sweet mystery of physics. As you will discover during future breakfasts, mass is also a ubiquitous property of an object that figures repeatedly in the laws of physics."

"We better not have any more eggs Benedict," Beth noted "unless we want to increase *our* masses unduly."

# Chapter Three:
# Breakfast of Apple-Gravity Pancakes

## *Why Does an Apple Fall From a Tree?*

"Before you start today's mini-lecture," Beth said the next morning just as I forked a delicious apple pancake into my mouth, "I've been puzzling over Newton's third law. How does mass figure in the fact that for every action there must be an equal and opposite reaction?"

"According to Newton, the mass of an object, when multiplied by its speed, determines its momentum. Suppose a ball strikes a hard surface like the floor. If the collision is one we call *elastic,* the ball will bounce back with the same speed with which it struck the floor. What actually happened to cause the ball to reverse its momentum? Just before hitting the floor, the ball had some momentum equal to its speed multiplied by its mass. When it struck the floor, it first had to decelerate to come to a stop at the floor. This change in momentum produced a force which was exerted on the floor by the ball. The third law then required the floor to exert an equal and opposite force on the ball. It is this force that caused the ball to come to a stop. All this is in full accord with Newton's third law which requires every action to produce an equal and opposite reaction."

"These two balancing forces explain what enabled the ball to come to a stop," Beth interjected, "but what became of the ball's initial momentum which, according to Descartes, had to be conserved?"

"The momentum of the ball was transferred to the floor in the collision. The floor, of course, has a nearly infinite mass when compared to that of the ball so that it gained an infinitesimally small speed, too small to be noticed. Yet the transfer of the floor's momentum back to the ball is what

21

caused it to regain its speed in the opposite direction. And were it not for friction and other losses, the ball could keep on bouncing forever, In fact, it is this kind of exchange that makes it possible for us to walk. At each downward step, the rigid floor exerts an equal and opposite force on the sole of the foot. By comparison, if you try walking through soft snow which can deform under the force of your weight, it takes a much larger effort to lift your feet up each time you continue walking."

"So even if we are not aware of it," Beth interposed, "Newton's laws help us at every step!"

"That's very true," I agreed. "Another example of how mass figures in the third law is illustrated by a rocket traveling in outer space. When the rocket's fuel ignites, it expels tiny bits of exhaust gases out of the rocket engine's rear (Fig. 3). The force necessary to do this, engenders an equal and opposite force acting on the rocket. If the particles are ejected to the left, then the reactive force acts on the rocket to the right. Even though each particle expelled has a very small mass compared to the mass of the rocket, when combined, these tiny masses multiplied by their accelerations to the left add up to a sizable total force acting on them to the left. The equal and opposite reactive force they produce then acts to accelerate the more massive rocket to the right."

"I thought it was the pressure from the rocket's exhaust pushing against the surrounding air that propelled the rocket," Beth commented.

"This is a commonly held misconception. In outer space there is no air to push against, yet it is possible to maneuver the rocket even out there. The way that can happen is fully explained by Newton's third law of motion. As long as we accept the intrinsic property of matter called mass, Newton

Momentum
of gas

Momentum
of rocket

Fig. 3.   The total momentum of the ejected exhaust gases acting on the left must equal the total momentum of the rocket to the right in order to conserve the momentum of the fuel — rocket system to the same value it had before the fuel's ignition.

showed how all aspects of motion can be interrelated in a consistent and logical matter."

"How did Newton finally connect up the weight of an object with the gravitational pull of the earth on it?" Beth now wanted to know.

"Once he had related the force acting on a body to the acceleration that it produced, Newton naturally began to speculate about what the various forces acting on a body might be. He knew from Galileo's very careful measurements that any freely falling object on earth underwent the same acceleration in a straight line toward the center of the earth. A force directed toward a center is called a *centripetal* force, by the way. I stress this because Newton also knew from the observations of others that there is at least one way to produce a centripetal force artificially and that is to revolve an object about a center. A familiar trick is to spin a bucket containing water rapidly overhead without spilling the water (Fig. 4).

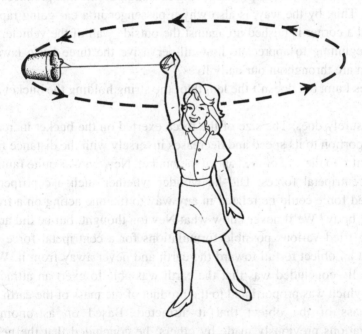

Fig. 4.   As the girl spins the bucket around her head, she exerts a centripetal force through the string tied to the bucket. Note that, if the bucket were not so attached, it would continue to travel in a straight line (indicated by the succession of arrows), which is exactly what would happen to it if the string should break.

The string tied to the bucket exerts a centripetal pull on the bucket and keeps it from continuing to travel in a straight line in accord with its inertia."

"Oh, I've already heard about this!" Beth exclaimed excitedly. "The string exerts a centripetal force on the bucket but the bucket exerts an equal and opposite *centrifugal* force on the string. So does the water inside the bucket and that is what keeps it from spilling out!"

"That is essentially correct. What actually prevents the water from spilling out is that its inertia makes it 'want' to keep moving in a straight line. It is the bucket's circular orbit that causes the bucket's bottom to exert a centripetal force on the water. If you've ever gone on one of the many different circular rides in an amusement park, then you know how your entire body is pressed against the outer wall of the spinning vehicle. People commonly refer to this as a centrifugal force when, in fact, it is your inertia trying to keep you moving in a straight line that produces this effect. This, by the way, is also why a passenger in a car going rapidly around a corner is pushed up against the outside wall of the vehicle. Are you beginning to appreciate how all-pervasive the three basic laws of motion are throughout our daily lives?"

"Yes I am, but doesn't the length of the string holding the bucket also matter?"

"It surely does. The size of the force exerted on the bucket increases in proportion to its speed and decreases inversely with the distance from the center to the end of the rope at the bucket. Newton was quite familiar with centripetal forces. Did he wonder whether such a centripetally directed force could be related in any way to the one acting on a freely falling body? We'll never know what Newton thought, but he did admit having tried various possible formulations for a centripetal force that caused an object to fall toward the earth and never away from it. What he finally concluded was that the earth was able to exert an attractive force which was proportional to the product of the mass of the earth and the mass of the object that it attracted. Based on astronomical observations previously made by others, he concluded that the nearly but not quite circular orbits of the planets 'circling' the sun in our solar system could be explained by a solar attractive force that decreased in proportion to the square of the distance to the planet."

"Is *that* the famous inverse-square law of Newton?" Beth asked. "I seem to have heard of it in other connections although I'm not sure I really understand it."

"Yes, that's what it's called. Let me illustrate how this inverse-square rule works. You know that, when you look through the view finder of a camera, you see a certain size area of view framed in it. If you back away from the object that you are photographing, the size of this area increases; if you move closer to your object, the amount of area that you can record on the film decreases. Since area is equal to distance squared, it is a simple exercise in geometry to prove that the area that your view finder sees is directly related to the square of the distance between you and the object you're viewing. Next, suppose that you replace the camera with a can of spray paint (Fig. 5). The further you move from the can, the larger the area sprayed becomes. Since the amount of paint delivered to this area is unchanged, the thickness of the paint decreases as the total area covered increases. This thickness is inversely related to the square of the distance from the spray can. Today we know many more examples of where such relationships exist, but at the time of Newton's pronouncement it caused quite a stir in the scientific community."

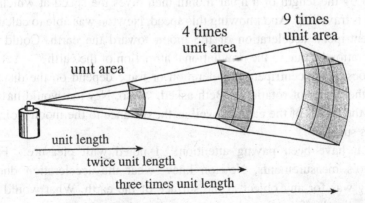

Fig. 5.   The spray paint reaching each of the three cross sections indicated must cover an area that increases with the square of their distance from the spray can. The thickness of paint that covered the area one unit distance away from the can must cover four times that area two unit distances away, etc. The thickness of the paint delivered, therefore, must decrease inversely with the square of the distance.

"Now I see what is meant by an inverse-square law." Beth observed. "That's a really good illustration."

"One of the things that concerned Newton very much," I continued, "was how the gravitational attraction between two bodies could be transmitted through space without any visible contact or other type of interaction between them. Descartes had argued previously that any interaction between two objects required some means for transmitting it. Clearly, a force can be transmitted by contact, for example, by pushing or pulling an object. Even sun light reaches the earth, Descartes argued, because some invisible medium between the earth and the sun enables the light to pass through it in the same way that sound can travel through air or a more solid medium but not through a vacuum. Nevertheless, everything seemed to point to the action of the attractive force between two bodies over a distance separating them rather than through some medium connecting them.

If gravitational attraction nevertheless exists and explains why an object falls toward the earth, Newton wondered, could it not also explain why the moon circles the earth in a regular orbit? The distance from the earth to the moon was already known so that it was a relatively simple matter to calculate the circumference of the moon's orbit. Dividing this length by the length of a lunar month then gives the speed at which the moon is travelling. And knowing this speed, Newton was able to calculate the centripetal acceleration of the moon toward the earth. Could this acceleration be due to the gravitational attraction of the earth?"

"Doesn't the centripetal acceleration or force depend on the distance from the center of rotation?" Beth asked. "If so, Newton would have to know the radius of the earth as well as the distance to the moon from the earth's surface."

"You have been paying attention!" I noted with pleasure. "From Galileo's measurements, Newton knew what the acceleration due to gravity was for any object at the surface of the earth. What would this acceleration be if the object's separation from the earth's center were to be increased by the distance to the moon? According to his law of gravitational attraction, Newton reasoned that it should be decreased by the square of the separation. When he carried out the actual calculation, the two centripetal accelerations turned out to have different values. After rechecking his calculations, the discrepancy remained. It finally caused Newton to set his theory aside and to concentrate his efforts on other studies.

A few years later, however, the British astronomer Edmund Halley persuaded Newton to repeat his calculations using a newly revised value for the radius of the earth. Voila! The two accelerations turned out to have the same value within about one percent! Encouraged now to make his calculation public, Newton went before the Royal Society of London to describe 'how the moon was falling toward the earth'. In explaining that the earth's gravitational pull was causing the moon to deviate from the straight-line path required by its inertia, he likened it to an apple falling from a tree. In this way, probably, the legend about Newton and his discovery of gravity by observing falling apples was born."

"I'm so glad that apples fall toward earth and not away from it." Beth interjected with a smile on her face. "So that I can enjoy them inside a delicious pancake."

"Just so. I want to emphasize, however, that the effect that Newton's revelation had was extremely dramatic. Not only had he succeeded in demonstrating the correctness of his gravitational law, but Newton had now shown that this law extended to the moon and the heavenly bodies above! For the next one hundred and more years, astronomers and mathematicians would carry out calculations of the motion of planets orbiting the sun due to its gravitational attraction. These analyses were eagerly followed by the educated society of the times and not just by other scientists. Particularly in France, which considered itself to be the cultural center of Europe, these arcane calculations enjoyed a popularity not matched even today by the far more startling scientific discoveries of the twentieth century."

"How could Newton make all these calculations?" Beth wanted to know. "Did he know what the mass of the earth or the moon was?"

"No, he did not know their masses. By comparing the centripetal acceleration at the surface of the earth to its calculated value at the height above the earth at which the moon orbits it, Newton didn't have to make use of their respective masses in his calculations. All he needed to know accurately was the ratio between the radius of the earth and the distance to the moon. In order to determine the masses, one would need to measure the force of gravitational attraction directly. Such a measurement was ultimately made about one hundred and fifty years later by Henry Cavendish. Possessed of a sense of humor no less than Newton's, Cavendish presented his results by purporting to show 'how to weigh the earth'."

"Was Newton able to figure out what causes the gravitational attraction that two objects feel for each other?" Beth asked now.

"No he did not. Nor did he know why this force decreased when their separation increased. Moreover, we don't know the answer to your question even today. We have better insights into exactly what goes on in outer space, of course, but what is responsible for the gravitational attraction remains a mystery to the present time. It comes down to this: if you accept the idea that every object has an invisible and untouchable mass, then its motion, as well as its attraction for another object also possessed of a mass, has to obey Newton's various laws. The acceptance of the concept of mass then enables us to explain correctly how all objects behave on earth. It also enables us to predict correctly such behaviors throughout the universe. Moreover, we can observe the correctness of these predictions with our telescopes and space exploration devices. Little wonder, therefore, that physicists accept the reality of mass with a faith that is as total as is the faith some others show in their religious beliefs or their horoscopes."

"You're not suggesting that astrology is based on verifiable truths?" Beth asked in mock horror.

"No, no, no!" I responded in a more serious tone. "Astrology is sheer fiction based entirely on an age-old desire to be able to know what lies ahead. Our various religions were established by people who were inspired to do so in different ways. They typically demand the acceptance of the existence of God without any further proof required or offered. Physicists would like to think that their laws are based on absolute truths because they can verify their correctness by direct observations of nature. It is nevertheless true that physics is based on certain assumptions that can't be proven so that they too have to be accepted on faith."

"Aren't you also implying that an atheist can no more prove that there is no God than the most devout believer can prove that a God exists?"

"Sure, believers and nonbelievers can argue endlessly, but they lack the means to prove the validity of either view." I added. "This may be the reason for the never ending strife among the followers of different religions. By comparison, one doesn't have to be a physicist to appreciate the correctness of Newton's law of motion."

"But how did you like your apple pancakes?"

"Can't you tell by observing that my plate is empty?"

# Chapter Four:
# Breakfast of Cereal and Calories

## *But It Takes Energy to Keep Moving*

"I have thought up a good example of Newton's third law for you," Beth announced the following morning "It's not possible to kiss someone on the lips without being kissed in return!"

"That's very good! But did you ever stop to consider that kissing someone requires the expenditure of energy? If not, you're in very good company. Neither Descartes nor Newton ever gave it a thought. In fact the term *energy* did not appear until a lecture delivered by Thomas Young before the Royal Institute of London in 1801."

"Is it true that Newton avoided romantic entanglements? Was that so that he could save up his energy?" Beth asked with a twinkle in her eyes.

"That I don't know, but let me put it to you in another way." I said. "We frequently refer to energy in our daily conversations, but do you know what energy really is?"

"Well the poets claim that energy is what makes the world go around," Beth continued mischievously, "but no, I really don't know what energy is."

"Again you're in good company. It is very difficult to define what energy is because, in reality, it is nothing but a mathematical construct. By this I mean that we can express energy in an algebraic formula and use it in many different ways to explain the workings of nature. But like the concept of mass, we can't see it or touch it. As is the case for mass, we can measure energy and, therefore, we can make excellent use of it. In fact, physics has been described often as the study of matter and energy, so you see that energy plays a very central role in the rest of our story."

"Tell me more. Where did the study of energy begin?" Beth wanted to know.

"I believe that Thomas Young used the term first, but he did not invent the concept which evolved over time from the studies by many different people. It has become popular to define energy as the ability to do work. For example, electricity is a form of energy because it can operate a motor; heat is a form of energy because it can cause water to boil. This can produce steam which, in turn, can drive a generator to produce electricity, and so forth. What this also illustrates is that energy can be changed from one form to another and, as became clear by the middle of the nineteenth century, total energy must be conserved! This is our second conservation law in nature and like the first, conservation of momentum, we can't prove it nor can we derive it from more fundamental truths so we, simply, have to accept it because it works so well in explaining natural phenomena."

"My mother likes to say that 'work is love made visible'." Beth commented. "Are love and energy somehow related?"

"Certainly 'making' love is a form of mechanical work that requires energy. But physicists require more precise definitions than those used in our daily conversations. They define work as a measure of the force applied multiplied by the distance through which this force acts. Without burdening ourselves with the details here, let me point out that such a definition enables us to measure the work done because it is relatively easy to measure a force and a distance. By *defining* energy as the equivalent of work, we then have a straightforward way of measuring energy. This is equally true of electric energy, or the energy arriving in the form of heat and light from the sun, or that consumed in generating the proverbial sweat on our brow. This is why the above definition of energy is both popular and useful. To paraphrase your mother, Beth, one can say that 'work is energy made visible'."

"As you doubtlessly know," I continued, "we also distinguish two kinds of energy: the *potential energy* that a body can store to do work later and the *kinetic energy* which it expends in the process of actually doing work. Thus a car battery can store electric potential energy and subsequently release it as the kinetic energy that can operate the electric lights or mechanical air conditioner in a car. Similarly, a dam can store water in an artificial lake located near the top of a mountain and then release it to form a waterfall that can rotate a wheel attached to an electrical generator. The potential energy of the stored water is expressed by its mass multiplied by its height above the wheel. The kinetic energy of the falling water is similarly defined by its mass multiplied by one-half of its speed squared. Again, it is not the algebraic relationship that matters but the fact that both kinds of energy include mass in their definitions. I stress this because we accept mass as a given of nature. We also accept energy as a given of nature. We accept these two concepts of physics entirely on faith because they are so all pervasive in our understanding of the nature of the world around us."

"Doesn't Einstein's famous equation, $E = mc^2$, relate energy and mass directly?" Beth asked.

"Five years into the twentieth century, Einstein published his first paper on relativity which deals with the motion of bodies and light throughout the universe. One of the important results of that analysis was the formula you cited. What it actually states is that mass and energy are *equivalent* to each other. Their magnitudes are related by the speed of light, $c^2$."

"Wasn't this involved somehow in the development of an atomic bomb?"

"The idea that mass and energy were really different manifestations of the same thing, whatever that 'thing' may be, was truly revolutionary. Carrying this concept to its logical conclusion, it suggested that an atom having a tiny but finite mass is not unlike a storage battery because its mass can be converted into kinetic energy. The first such direct conversion of the mass of uranium atoms into energy took place under the seats of the University of Chicago's football stadium, where it was used primarily to generate some heat. Tragically, several years and several billion dollars later, the same mass-to-energy conversion was used to devastate much of two Japanese cities as a prelude to Japan's acceptance of the terms of unconditional surrender demanded by the victorious allies in World War Two."

"Yet, like so many other scientific discoveries, atomic energy can be used to benefit humankind as well. To return to our topic, tell me more about how energy and heat are related," Beth requested.

"First let me tell you what heat is. A Scotchman by name of John Black pictured heat to be a liquid that could flow into a body to warm it and out of it to cool it. He named this liquid calor after the Latin word for heat.

Being a scientist, Black defined a basic unit of heat as the amount of heat needed to raise the temperature of one pound of water by one degree Fahrenheit. This notion was accepted by a French chemist, Antoine Lavoisier, who renamed the fluid *caloric*. Nearly one century had to pass before it was realized that heat was not some form of fluid but instead was a form of work. This happened when the curiosity of a professional soldier was aroused while he was observing the inadvertent heating of cannon barrels being bored for his master."

"Sounds like an interesting story," Beth interjected. "Can you tell me more?"

"Some years after serving as a spy for the British during the Revolutionary War, a renegade American, named Benjamin Thompson at birth, wound up as an administrator in the court of the Elector of Bavaria in Europe. The Elector rewarded Thompson for his loyal services by granting him a peerage for which Thompson chose the name of his village of birth in America and so became the first Count of Rumford in 1790. Eight years later, while supervising the boring of cannon barrels, the Count noticed that they became extremely hot, particularly as the boring tool became progressively duller. How could the wearing away of the boring tool be caused by a calor flowing into the barrel? Unable to explain the heat generated by the boring in terms of the prevailing caloric theory, Count Rumford made the bold assumption that the heat was generated by the friction of the boring tool against the barrel (Fig. 6). This is not unlike the way we can heat our hands by rubbing them against each other on a chilly day."

"Did Rumford conduct any actual experiments or was he satisfied with having thought up a new way to describe heat generation?" Beth wondered aloud.

"Actually he did perform experiments and attempted to establish the energy equivalent of mechanical work but his calculations contained some errors that prevented an agreement with his measured values. Forty-four years later, a German physician named Julius Mayer corrected these calculations but, not being a physicist, his contributions were ignored by the scientific community of that time. It was left to the self-educated son of a brewer of beer, who decided to carry out extremely careful measurement, to establish the equivalence of the heat generated to the

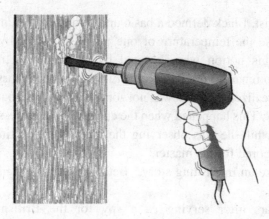

Fig. 6.   The drill bit and the wood chips produced in the panel being drilled become heated by friction with the rapidly turning drill. This illustrates how the mechanical energy of the drill is transformed into thermal energy. But can you *see* or *feel* the energy?

mechanical work producing it. James Prescott Joule performed literally thousands of experiments to determine this quantity precisely. Nevertheless, he had great difficulty in having his results accepted by the Royal Society until the highly regarded British physicist William Thompson (later to become Lord Kelvin) expressed his support. By 1850, however, the contributions of Joule were fully recognized and he was elected to membership in the Royal Society. In fact, he was awarded its coveted Copley medal sixteen years after that."

"By the way," I added, "the transformation of mechanical energy into heat is so all pervasive that no mechanical motion is entirely free of it. This is self evident when friction is present, but it even takes place when invisibly tiny electrons flow through a conducting wire. It also occurs, moreover, when the kinetic energy of a falling rock is converted into heat upon hitting the ground below. If you don't believe that such impacts generate heat, try touching the head of a nail after striking it repeatedly with a hammer!"

# Chapter Five:
# Breakfast of Hot Cakes
# with Energy

## How Hot is It?

"You said that the caloric theory of heat acting like a fluid was discredited," Beth observed as she poured warm syrup over a stack of hot cakes. "But aren't calories used to measure heat?"

"We do indeed express heat or *thermal energy* in units of *calories* but that's just a carryover of the Latin term for heat and has nothing to do with heat as a kind of fluid. As a matter of fact, a substance absorbs thermal energy by increasing the kinetic energy of the atoms comprising it. Maybe we can talk about this some other time, if we stick to having these breakfast talks long enough. Anyway, the transfer of heat from a warm to a colder substance is a form of energy transfer just like the friction produced by rubbing transforms the mechanical energy into heat."

"I see!" Beth exclaimed. "When I warmed the syrup in the pot, the molecules gained kinetic energy from the electric burner and the syrup became hotter. So, is temperature also a measure of heat?"

"By convention, a calorie is the appropriate unit for measuring thermal energy," I replied. "The *effect* that adding or substracting heat has on a body is to raise or lower its *temperature*. It is important not to confuse these two attributes of heat."

"Let's see if I got it straight," Beth asked. "If I add heat to the syrup in a pot, I increase the number of calories it contains, and that raises its temperature. Then should I dump some ice cubes in it, I would be adding cold to the syrup so that its temperature should drop. Am I adding negative calories when I add the ice?"

35

"No, because there are no negative calories. Heat is thermal energy, but cold is, simply, the absence of heat. Just like a balloon, which can be filled with helium gas atoms, or any gas it contains can be evacuated, and the vacuum produced contains no atoms of any kind. We don't need to have 'negative' atoms to produce a vacuum and we don't need negative calories to produce cold. We simply remove the thermal energy from a body and its temperature drops as a result."

"So, you can't add cold, you can only remove heat from a body," Beth murmured. "What about temperature? I know that there are at least two scales for measuring heat: Fahrenheit and centigrade. Why does the Fahrenheit scale begin with 32 degrees while the other starts with zero?"

"According to one legend, true or not, the German instrument maker Gabriel Fahrenheit chose the temperature recorded on the coldest day in his native Free City of Danzig to fix the point on his thermometer as zero. Then he chose the body temperature indicated by his sick wife one day to fix another point at 100 degrees. The idea of having two fixed points separated by 100 divisions or degrees is definitely attributed to Fahrenheit. He also was the first to hit upon the idea to use the temperature of a mixture of water and ice and that of boiling water as reference points since they were easier to reproduce consistently. On Fahrenheit's thermometer, these two points turned out to be separated by 180 degrees. All those using this scale, therefore, are stuck with these seemingly arbitrary numbers."

"Later on," I continued, "the Swedish astronomer Anders Celsius suggested that it would make life much simpler if we were to set the temperature of ice water as zero and that of boiling water as one hundred. Thus was born the Celsius or centigrade scale. All scientists and virtually all countries except the United States have adopted the Celsius scale. Hopefully it will become unanimous some day."

"Let's see," Beth concluded, "when I heat ice to 32 degrees on the Fahrenheit scale, it turns into water. If I add more calories to it, until it reaches 212 degrees Fahrenheit, it will then boil and turns into steam. Is that right?"

"Almost right," I responded. "What happens when you raise the temperature of ice to zero degrees centigrade or 32 degrees Fahrenheit is that the ice becomes ready to melt. Actually the melting process requires that the water molecules frozen rigid in the ice gain enough kinetic energy to break loose to form a liquid. We call this the *latent heat of melting* and it represents an amount of thermal energy that is absorbed by the ice *without raising its temperature*! In other words, the ice water stays at the same temperature of 32 degrees Fahrenheit while all the ice converts into water. Only then can the water's temperature start to increase due to the extra heat being added to it.

The same thing happens at the boiling point. An amount of thermal energy called the *latent heat of vaporization* must be added without raising the water temperature above 212 degrees Fahrenheit before the water molecules gain enough energy to escape as steam. What you are witnessing at both points is the energy required to change what is called the *state* of a substance."

"Aha!" Beth exclaimed. "That's why I seem to remember that there are three states of matters: gas, liquid, and solid. Right?"

"Right you are! But we can talk about these three states some other day."

"From my reading about superconductivity in the newspaper I seem to remember a reference to an absolute scale of temperature," Beth observed. "Is that the centigrade scale that scientists use?"

"No. The *absolute* scale of temperature was established much later. It came about from the study of how gases are affected by changes in the temperature, pressure, and volume of the gas. Such studies by chemists and physicists made it apparent that a zero point for atomic motion occurs 273 degrees on the centigrade scale below the freezing point of water. At this incredibly cold temperature, their kinetic energy drops to zero and their motion ceases for all practical purposes. Physicists find it convenient, therefore, to refer to this temperature as the absolute zero of temperature, since nothing can be colder than this. Using the same degree divisions as those on the Celsius scale, the resulting absolute scale puts the ice-water temperature at 273 degrees while water boils at 373 degrees. It is also known as the *Kelvin* temperature scale after Lord Kelvin, who did much to popularize this concept."

## Putting Heat to Work

"There's something I've always wondered about steam." Beth noted. "Why does boiling water make the lid of a pot clatter?"

"When water boils, it undergoes a change of state to steam. As the steam is heated further, it exerts an increasing pressure on the lid until the pressure is large enough to raise the lid. This allows some of the compressed steam to escape. The steam remaining then can expand. This decreases the temperature of the remaining steam and lowers its pressure so that the lid bangs shut, only to start the process all over again. Popular legend has it that such observations inspired James Watt to invent the steam engine. In actual fact, a steam engine was first developed by a British iron monger, Thomas Newcomen, who used it to pump water out of the tin mines in Cornwall near the start of the eighteenth century. Newcomen's steam pumps did not work very well so that they never really caught on. Watt was working at that time as an instrument maker in a small college in Scotland. Someone gave him a Newcomen pump to repair and he couldn't resist trying to improve its efficiency."

"The basic steam engine is a very simple device." I continued. "If you boil water in an old-fashioned kettle (Fig. 7), the steam produced escapes through the spout once the water starts to boil. If you place a fan or a paddle wheel in the path of the expanding steam, it will rotate the fan or wheel. By attaching some device to the rotating wheel you can get it to perform work, like pumping water or generating electricity, or whatever.

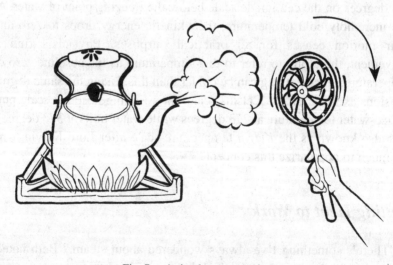

Fig. 7.　A simple steam engine.

By the way, don't *you* try putting things in the path of the steam escaping from the kettle because it is very hot and can burn you fingers badly."

"It doesn't seem to be a very efficient device, however." Beth observed. "The steam escaping from the kettle causes the water level to drop so it needs constant replacement."

"Quite true," I agreed. "By enclosing the steam path somehow, the steam passing over the paddle wheel can be collected in a condenser where it cools and liquifies so that it can then be recirculated back to the boiler. By trial and error, Watt discovered that the pump worked better when the chamber in which the steam was generated was insulated so as to retain as much of the heat as possible. The expanding steam was used to drive the piston in the pump that raised the water. So, by increasing the heat in the expansion chamber, more mechanical energy could be delivered to the piston rod."

"That was clever of him, eh what?" Beth loved her little puns.

"Watt's biggest discovery," I tried not to respond, "came after he realized that he could make the pump even more efficient by removing the expanded steam to a separate chamber. There it would reliquify by letting the remaining heat escape to the environment. In Watt's steam engine, the condensed water is then pumped back to the well-insulated expansion chamber where it is again heated into steam. This steam drives a turbine that can be connected to any machine, not just the pump. With such variations on Newcomen's theme, James Watt helped launch the industrial revolution in England."

"How efficient is a Watt steam engine?" Beth wanted to know.

"Efficiency in physics expresses how completely energy is converted into work. Stating it as a percentage, one measures the useful work done and divides it by the total energy consumed in producing that work. In addition to the energy put into heating the water, like any good accountant, one must add all of the losses, including friction, work done in pumping the water back to the steam chamber, any heat lost to the environment, etc., to account for all the energy that went into things other than doing useful work. The ratio of useful output to the total input then will turn out to be no more than one half so that even the best steam engine is only about 50% efficient. Watt, who was a better instrument maker and business man than scientist or engineer, did not attempt such theoretical analyses. Instead, he continued to improve his engines by trial and error and his personal well being by patenting everything he built."

"Wasn't launching the industrial revolution in England enough? Did he also have to theorize about what he was doing?" Beth wondered aloud. "By the way, who *did* figure out how a steam engine really works?"

"The actual analysis of what takes place in such a process was undertaken by a Frenchman, Sadi Carnot, trained as an engineer but thinking like a physicist. Serving as an officer in the French army during the first part of the nineteenth century, Carnot was fully aware of the practical limitations of the Watt steam engine. Yet he wondered, how much better could an ideal engine perform if it had no steam seeping out of it or friction to slow it down? Such an ideal engine would have one unique distinction from all real engines. Since an ideal engine suffers no losses, the useful work that it produces should just equal the total energy or work put into driving the engine. In other words, such an operating cycle can work in either direction."

"It sounds to me like you are describing a perpetual motion machine," Beth observed.

"That's very true," I responded, "but keep in mind that this is an ideal not a real engine that Carnot was analyzing. As we shall see, it is this analysis of an ideal engine that shows why a perpetual motion engine is, in fact, not possible."

"The study of the flow of energy," I continued, "especially in the form of heat, is so important that it has become a separate subfield of physics called *thermodynamics*. As is true of most laws of physics, the laws of thermodynamics are also quite simple. What tends to complicate them is the need to account fully for every possible kind of energy transfer taking place in a physical process. Basically this is like balancing one's cheque book each month or preparing a profit-and-loss statement each quarter for an ongoing business. Accountants must repress any urges to show a positive balance at the end of the accounting period by masking any losses suffered by one of their clients Similarly, scientists must be careful not to overlook any energy losses in the system under their analysis. To help them do this most effectively, fairly sophisticated mathematics were developed toward the end of the nineteenth century. In fact, chemists, engineers, and physicists involved in this process have each developed mathematical languages that are somewhat unique to their fields. That's why it is not unusual to find an introductory course in thermodynamics being offered by each engineering department, as well as the chemistry and physics departments of the same university."

"It must be pretty important for all those fields," commented Beth. "Did Carnot's calculations lead to anything practical or were they all conjectural only?"

"To answer your question, let's take a look at Carnot's analysis of the ideal or fully reversible engine. In a greatly simplified version of his laborious and careful analyses, think of the energy of the hot steam in the boiler as the potential energy at the start of an operating cycle. As the steam passes over the turbine wheel (Fig. 8), it converts its kinetic energy into the useful work of turning the wheel. The expanded and cooled steam then goes into a condenser before being pumped back to the boiler where the cycle is started anew. The temperatures of the steam before it enters the turbine and after it emerges from it, therefore, can be used to represent, respectively, the state of the system's energy before the steam did work and what it becomes after doing that work. It turns out that, provided these temperatures are expressed on the absolute or Kelvin scale, the difference

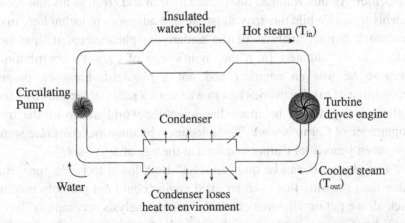

Fig. 8. In a simple steam engine, water is heated in a boiler, from which the steam produced is directed to a turbine whose wheel is turned by the steam as it expands and cools. The steam is further cooled and liquified in the condenser from where the water is pumped back to the boiler.

between the input temperature and the output temperature, divided by the input temperature, yields a ratio that is a measure of the efficiency of this cycle. The heat given off to produce the temperature drop represents the work done by the turbine."

"It seems to me," Beth observed, "that all one has to do is to decrease the output temperature to zero and then the difference in the numerator will be the same as the denominator so that we would get 100% efficiency! Right?"

"Your analysis is perfect, but keep in mind that the temperatures are on the absolute scale and it is not possible to get down to absolute zero in practice." I explained "Moreover, the steam will have already turned into water at 373 degrees."

"You do have a way of throwing cold water on some of my best idea!" Beth said jokingly.

"Called the Carnot efficiency," I resumed, "the ratio representing output-to-input energy simply shows that the input temperature must be larger always than the output temperature. This way the efficiency is positive and a negative efficiency has no physical meaning. Moreover, it follows that the best way to increase the efficiency is to raise the temperature at the input and reduce it as much as possible in the condenser. All this Watt had discovered by trial and error, as his numerous patents testify. While this may illustrate that advances in technology may precede a correct analysis of the underlying phenomena, it does not detract in the slightest from the significance of Carnot's contribution. Because he was an engineer and not a physicist, however, proper recognition of his discoveries had to wait until a series of lectures by Lord Kelvin during which he made the scientific world aware of the true importance of Carnot's work. These lectures, by the way, took place some seventeen years after Carnot had died at the age of thirty-six."

"You physicists can be quite snooty!" Beth loved to tweak me. "But other than showing that even an ideal engine could not possibly give us back all we put into it, what else did Carnot's analysis accomplish?"

"What Carnot's analysis of the ideal heat engine showed was what happens when thermal energy is converted into mechanical work. Since the Carnot efficiency of even an ideal engine is of necessity less than 100%, it follows that any real engine must have an efficiency that will be even smaller. Why this has to be the case cannot be explained by any other basic law of physics. This is why the laws of thermodynamics take on a significance like that accorded to various conservation laws, the concept of mass, and the other facts of nature that we can describe and utilize, but not explain beyond their expression by mathematical equations. It is in the interpretation of Carnot's results that Lord Kelvin made some important contributions, as did other scientists that followed."

"Our remaining hot cakes are cold and so is the syrup." Beth observed. "If you don't want me to reheat them, I'm ready to return to my work. Maybe this weekend, I'll make you some French toast out of French bread in honor of Carnot's memory."

# Chapter Six:
# Breakfast of French Toast

## *First Law: You Can't Win*

"Here's the French toast I promised." Beth announced at the start of our next breakfast.

"My favorite!" I exclaimed.

"I'm glad. But you promised to tell me more about Carnot and Lord Kelvin."

"As we've seen, the Carnot cycle demonstrates that it is not possible to have a perpetual motion machine. Even in an ideal heat engine, which suffers no losses of any kind, we can never convert all of the input energy into an equal amount of output energy or, more precisely, into an equal amount of work. In trying to understand why this should be, Lord Kelvin and others developed what has now become known as the second law of thermodynamics. It may be an oddity of physics but, what we call the first law of thermodynamics was not actually accepted by physicists as a general law until many years later. Probably because it now seems that the first law *should* have been discovered first, it has been given such precedence in all text books on this subject. It is also true that it is much easier to understand the second law and to derive the necessary mathematics if we can start with a knowledge of the first law. Carnot, of course, didn't have the benefit of such knowledge."

"All right already," Beth couldn't wait, "what *is* the first law?"

"What it says is that the total energy of any system is the sum of all the component energies making up that system plus any work that may be done by or on the system. The reason why it took so long to become adopted is that the first law is so all encompassing."

"Well, what's the big deal? Doesn't that just mean that the total is a sum of all the parts?"

"Absolutely! Now, however, expand that concept to include the entire universe. If the total energy of the universe is a sum of all the various energies in it, can that total ever change? The answer is that it can't. The only thing that can happen is that electrical energy can be transformed into thermal energy, and the thermal energy into mechanical energy, and so forth, but the total remains the same. In other words, the total energy in the universe must be conserved."

"Yes, but what about the losses in any energy conversion process?" Beth wondered aloud. "Didn't Carnot just prove that even an ideal engine will not be 100% efficient?"

"What is lost is some form of energy, but where does it go? It doesn't just vanish. In the Watt steam engine, for example, any frictional losses are converted into heat. This heat plus the heat losses in the original boiler and in the condenser which cools the steam down to water, all must be absorbed by the environment in some way. So the total energy in the universe is unchanged even though the steam engine is giving off thermal energy all the time. Any difficulty you may have in accepting that this is the way things really are, and have to be, should help you appreciate why it took scientists so long to agree that this conservation law accurately and reliably states how nature operates.

What's more, to be a really meaningful statement, the law of energy conservation must be quite general and cover all kinds of systems. For example, suppose the system is moving. Then we must include a kinetic energy term in the summation to maintain a proper balance. If other changes to the energy of the system should occur, it must be equally possible to include all of them without upsetting the validity of the first law of thermodynamics. All this turns out to be doable. Even after Einstein demonstrated that energy and mass were equivalent, so that both must be conserved throughout the universe, the law of energy conservation still held. For example, when nuclear power generation is analyzed, it is sufficient to include the equivalency of mass and energy, $E = mc^2$, in the accounting process. In fact, it is difficult to exaggerate just how broadly this basic law can be applied to all kinds of systems."

"All the *ad hoc* changes introduced in your examples, don't they make the first law somewhat arbitrary for what you've called a fundamental law in physics?"

"Not at all," I replied. "The reason the first law works so splendidly in any and all applications is that it is a purely mathematical construct that requires only mathematics to make it work. The system under consideration can be anything: a heat engine or a lump of coal, whose stored chemical energy is convertible into heat once the coal is ignited, et cetera, et cetera.

Conservation of energy *is* a fundamental law of physics because in all of our experience, with all kinds of systems, on earth and throughout the universe, we have not found a single violation of this law! We also have such absolute faith in it because it enables us to predict correctly any number of other events."

"It seems to me," Beth now observed, "that what the first law of thermodynamics really states is that you can never get more out than what you put in at the beginning. In other words, you can't win!"

"That's very well put, Beth, but even though the first law explains so much, it doesn't explains it all ..."

## Second Law: Nor Can You Break Even

"As useful as the first law of thermodynamics may be in describing how energy can be changed from one form to another, it totally fails to explain why the changes take place as they do. Take, for example, the case of heat flow that we discussed before. We know that heat only flows from a warmer body to a colder one, never in the opposite direction. The first law would be equally satisfied by heat flowing in either direction, just as long as the total amount of heat remained constant. Similarly, we know that the mechanical work of friction can be transformed into heat, but there is no direct way to reverse that process to recover the heat in the form of mechanical work directly. Again, the first law doesn't forbid it but our experience tells us that processes involving heat and work are not reversible."

"Isn't that the real meaning of the Carnot cycle?" Beth spoke up. "But, in the case of heat flow not involving work, how can we be sure that heat only flows from warm to cold and never backwards?"

"Right again! Carnot *was* the first to demonstrate that there is no way to have a real reversible engine. As to your second question, all of our observations of nature show that heat flows in one direction only.

Take the case of iced drinks You put ice cubes into a liquid so that heat will flow from the liquid into the colder ice and melt it in the process. If it were possible for the heat to flow in the reverse direction then, conceivably, the ice could get colder while the drink becomes warmer. Do you consider that a likely prospect?"

"No, I don't think it is very likely," Beth answered, "but how can we be absolutely sure that an ice cube will never be chilled by a liquid?"

"Well, of course, if you put an ice cube at, say, −10 degrees centigrade, into liquid nitrogen, which boils at −92 degrees centigrade, then the warmer ice cube *will* lose heat to the liquid. The fact is that we have never observed but one direction of heat flow. Had the discovery of the first law occurred first, one might say that the second law of thermodynamics would have had to be invented in order to explain such observations since the first law couldn't.

The progress of science, however, is rarely quite so orderly. As already noted, the second law was devised to explain why it is not possible to transform thermal energy into mechanical work without also cooling the medium that supplies the heat. Why this has to be I shall discuss shortly; but let me first point out a very dire consequence predicted by the second law. If you accept the statement that heat is lost in the process of doing useful work, then it means that gradually, *very* gradually to be sure, but irretrievably, the entire universe must be getting colder! Carrying this to its logical conclusion, in the very, very distant future, there must come a day when there will be no heat left to do any work!"

"And they call economics the dismal science!" Beth exclaimed "I think they should call thermodynamics that."

"And I would like to know why anyone would think that economics is any kind of science," I quipped in response.

"Whatever you may think of the predictions, those made by thermodynamics are always accurate and reliable. It may be based entirely on mathematical constructs, but thermodynamics describes unfailingly what happens whenever energy transfer take place. Can one make such a claim for economics?"

"I think we may be disgressing," Beth quietly noted. "Let's get back to the second law."

"Like most basic laws of physics, the formulation of the second law is really quite simple. It states that the energy that is available in any system to do work must equal the initial or internal energy of the system, less the change that takes place in a quantity called *entropy*, multiplied by the absolute temperature of that system. As long as the product of the change in entropy and the temperature is smaller than the internal energy, their difference will be a positive quantity and the system, therefore, has energy available to do work. Temperature expressed on the absolute scale has to be always positive, that is, it is larger than absolute zero. As long as the change in the entropy is positive also, then the externally available energy is less than the internal energy of the system and the first law is satisfied."

"Why can't the entropy be negative?" Beth was puzzled.

"If the change in entropy were negative," I replied, "we would be subtracting a negative product from the positive internal energy and the two minus signs would give us a plus sign, meaning that we would be adding the entropy-times-temperature term to the internal energy. This would give an available energy that had become larger than the starting energy of the system, in direct violation of the first law of thermodynamics. Since this can't happen, it follows that the change in entropy must always be positive."

"I think I smell a rat," Beth interjected. "The change in entropy, whatever that is, must always be positive, right? Is that why heat can only flow in one direction? But how could Carnot know that if, as you said, he didn't know about the law of energy conservation?"

"Yes, yes, and maybe he didn't." I replied. "It is unlikely that Carnot believed that his analysis of the operation of an ideal heat engine would explain the directionality of heat flow in all possible systems. This extension as well as the perfection of the notion of entropy came later and was the work of many others. As to what exactly entropy is, I can give you a mathematical expression that defines it but the word picture that comes to my mind is one of *disorder*. It turns out to be very useful to think of entropy as a measure of the amount of disorder that is present in any system. The change in entropy in the second law then becomes a measure of any change in the amount of disorder that takes place as a result of the external work done. Since the change in entropy must always be positive, however, this means that the amount of disorder in the world must be constantly increasing!"

"I don't think that I like these laws of thermodynamics," Beth quipped, adding facetiously, "I may even write my congresswoman and ask her to repeal them. But before I do that can you give me some examples of these changes in entropy?"

"Let's consider an ice cube. The water molecules in ice are arrayed in a very orderly manner, not unlike the tiles on a bathroom wall. When the ice is heated, the molecules increase their kinetic energy progressively by vibrating back and forth until, at the melting point, they start to break loose and the solid ice turns into liquid water. As previously noted, the temperature of the ice–water mixture stays the same while the ice absorbs thermal energy in breaking up its orderly molecular array. Clearly, we have increased the disorder among water molecules by heating them."

"I accept that," Beth acknowledged, "but don't we reorder the water molecules when we freeze the water and turn it into ice?"

"We *can* reorder the water molecules by putting them, say, into a refrigerator of some kind. But the operation of the refrigerator consumes more energy and creates more heat which creates even more disorder."

"Now I know I don't like these laws. The first law stated that we can't get something for nothing — that's discouraging enough — but the second law states that we shall always wind up with less than what we had at the outset. Do these laws have any redeeming virtues?" Beth finally asked.

"I don't know whether you will consider my next example redemptive, but one interesting consequence of ever increasing entropy in the world is that it serves as a measure of elapsed time. Consider a glass vase falling on a hard floor and breaking into a multitude of glass fragments. In terms of the glass vase, the disorder and, therefore, the entropy has increased. But suppose this break up was recorded by a cam corder. You could then run the tape backwards and watch the glass fragments reassembling themselves into the whole vase. The same would be true of any process in which we can record visually the increase in entropy with the passage of time. In this way, therefore, changing entropy measures the direction in which time moves so that it is sometimes called the arrow of time."

"I don't find this at all reassuring," Beth commented. "I am quite content to measure the progress of time using a watch and a calendar, thank you. Is there any more to this story or can we go on to consider some other aspect of physics that is less depressing?"

"There is a great deal more, of course, because the laws of thermodynamics are used repeatedly by engineers in evaluating our environmental problems and in seeking ways to alleviate them, or to optimize various processes to conserve costly energy consumption and waste. There is also one footnote that you may find interesting.

Early in the present century, a German physicist named Walther Nernst put forward a hypothesis for optimizing certain chemical processes by letting them drop to the absolute zero of temperature. Although one can't actually get to absolute zero, this analysis found almost immediate application in several industrial processes and helped Nernst win the Nobel Prize in chemistry. In fact, some people consider this to be a third law of thermodynamics. An English physicist, David Langford, subsequently observed that whereas the first two laws say that you can't win or even break even, this third law says that you can't stay out of the game either."

"I have one last comment to make about all this," Beth said. "Although the predictions of thermodynamics may be pessimistic, I was able to follow your description of them. Why then is thermodynamics considered to be such a difficult subject even by the scientists and engineers who make use of it?"

"The reason is that the mathematics can become extremely complex. There is no way that I could give you a simple example of this complexity, but let me try an analogy. As you know, it is possible to evaluate the profitability of any financial venture, large or small, by totaling up the dollar value of all sales and subtracting from that figure the total cost of producing them. As in thermodynamics, care must be taken to total all the costs, from those of the starting materials, to the commissions paid to the salesforce. The difference in these two totals is the annual profit and the ratio of the profit to the input costs represents the profitability for that year. This profitability is an important measure of any company's performance and should be scrutinized closely before deciding whether to purchase that company's stock. Analysis of the bottom line completely masks the work that went into producing the figures and tells you nothing about the complexity of the corporation itself. You might say that I have shown you the 'bottom lines' of the physics that we discuss at our breakfasts. I plan to continue doing that, with your approval."

"There is one last point I would like to make," I added. "Although looking at the profitability of a business venture may be similar to determining the Carnot efficiency of a heat engine, there is one very basic difference. There are no laws of economics or whatever to predict reliably the outcome or the profitability of a business venture. That is why some businesses do very well and others become bankrupt. When it comes to thermodynamic analyses, on the other hand, the results can be predicted always with accuracy and complete reliability."

I don't recall her exact rejoinder, but Beth murmured something like: "this may be what makes physicists sometimes sound so arrogant."

I would have preferred her to say that physicists are extremely confident but then why not let her have the last word?

# Chapter Seven:
# Breakfast of Cold Cuts

## *Go Fly a Kite!*

"The electricity went out as I started to toast your bread." Beth announced the following morning.

"What about the coffee?" I asked.

"That's O.K. It finished perking just before the power failure. But, it is such a nuisance to be without electricity," Beth frowned. "I guess we'll have to settle for cold cuts this morning."

"We can pretend that we are in Holland. Remember how they would serve us cold cuts for breakfast when we were in Amsterdam?"

"I do." Beth responded. "By the way, you were going to tell me about electricity and this power failure seems like a most propitious opportunity."

"Do you realize," I was glad to oblige, "that only one century ago electricity was known to just a few people? Two hundred years ago, electric currents were totally unknown. Even today there are many parts of the world, including some remote corners of the USA, where electricity's only presence is in the glimmer of a single light bulb."

"You're right!" Beth exclaimed. "I sure am glad it came along. I do remember the old ice box we used before we got our very first refrigerator."

"Well," I began, "let me tell you about how electricity was first observed. About forty-two years before Newton was born, an English physician in the court of Queen Elizabeth the First published a book in which he compared the strange properties of two kinds of minerals found in many places in the world. Called, respectively, *lodestone* and *amber*, they had intrigued observers for more than two-thousand years, but

William Gilbert was the first to document the differences in the way that these minerals behaved. It was also Gilbert who first demonstrated that lodestones could attract iron filings and each other. If allowed to float in a liquid, the lodestones invariably aligned themselves so that one of their ends pointed toward the south and the other toward the north. This fact was already known to Chinese and Arab navigators since the eleventh century but it was Gilbert who showed that the north-seeking end of one lodestone attracts the south-seeking end of a neighboring lodestone. He called this phenomenon magnetic *coition*, an appropriate term for a physician to choose."

"I like that term 'coition'," Beth giggled, "but what does it have to do with electricity?"

"As we shall see later, magnetism is intimately intertwined with electricity," I explained, "but it was sheer happenstance that Dr. Gilbert was describing the properties of lodestones and ambers in the same book.

As to amber, Gilbert reported that it could attract chaff after being rubbed vigorously but otherwise he didn't see any similarities between its properties and those of lodestones. In fact, he scolded his fellow physicians for attributing mysterious healing powers to both minerals without any proof of such a relationship."

"By the middle of the seventeenth century," I continued, "it was established that rubbing amber with a dry cloth causes it to attract small bits of various materials. Adopting the Greek word for amber, *electron*, this phenomenon became known as electric attraction. A member of the Royal Society of London reported that these electric properties could be transmitted through certain materials, notably metals, but not through others. It was the French Academician Charles Francois de Cisternay du Fay, however, who proposed that two kinds of electric 'fluids' were possible. One he produced by rubbing glass, precious stones, or porcelain bodies and so he called it *vitreous* electricity. The other he produced by rubbing certain resins, including the fossilized resin called amber, so that he named it *resinous* electricity. In 1734, du Fay also pointed out that bodies having like electricity repelled each other while those having unlike electricity attracted each other."

"While you were talking," Beth observed, "I rubbed the amber pendant you bought me two years ago on our cruise in the Baltic and, sure enough, it attracted to itself two little scraps of paper. Once they touched the amber, however, they were repelled by it. What causes that?"

"The electrically charged amber produces a like charge in the paper bits it touches." I explained "As du Fay had reported, like charges repel each other, so that after assuming the amber's charge, the paper now repels the like-charge amber."

"What fun!" Beth exclaimed. "Am I making electricity when I rub the amber pendant?"

"Yes. What you are making," I resumed, "we now call *static electricity* and its study became very popular during the eighteenth century, in part to satisfy the scientific curiosity of investigators and in part to entertain or, literally, shock one's guests. A German professor would stand a female student atop a hidden generator of static electricity and invite his male students to try kissing her. Any brave volunteer would be knocked back by a large unpleasant shock. What the young woman thought of her mysterious powers has not been recorded."

"Is this static electricity related to what happens when one walks across a rug on a dry winter day?" Beth now asked. "If I then reach for a metal door knob, I can get a pretty nasty shock."

"As you walk across the rug, you become electrically charged by rubbing your shoes against the carpet." I responded. "The term electric *charge* came into use to describe the condition wherein a quantity of electricity could be added to or subtracted from a body to render it electrically charged. This notion, incidentally, originated with the American inventor, printer, natural philosopher, patriot, and statesman, Benjamin Franklin. Unlike du Fay, Franklin realized that, what he called an 'electric fire', could be added to or subtracted from a body in the course of 'electrocising' it. This became known as the one-fluid theory of electricity and, according to Franklin, a body became either electrically *positive* or *negative* as a result. To demonstrate the validity of his theory, Franklin actually flew a kite on a rainy day in Philadelphia in 1752 and drew 'electric fire' from the storm clouds present. He survived to tell the tale because he had the prescience not to hold on to the wet kite string, but instead let a large key at the end of the string conduct the electricity to the ground below it. Being a very practical man as well as a keen observer, Franklin made immediate use of his discovery by inventing the lightning rod."

"Is that what lightning is — a discharge of static electricity?" Beth wanted to know. "How does it get into the storm cloud in the first place?"

"Storm clouds are collections of water droplets, snow flakes, and ice particles. As the cloud moves through the atmosphere, these particles in the cloud rub against the atmosphere causing the top of a cloud to become positively charged and the bottom to become negatively charged. As we now know, the negative charges are carried by electrons, which tend to accumulate in the negative underside of a cloud. This concentration of negative charges induces a positive charge in the ground below by repelling the negative electrons present on the surface of the ground. The electrons accumulated in the cloud can pass through the wet air to some positively charged object on the ground in what we see as a lightning flash (Fig. 9). This is why people are advised to avoid being near tall trees or in open expanses during a lightning storm."

"Oppositely charged bodies must exert very large forces," Beth noted. "A lightning bolt certainly can do lots of damage to whatever it strikes."

"You bet! The force exerted by two opposite charges on each other is considerable. In fact, it is one of the strongest forces we know.

A French engineer, Charles Augustine de Coulomb, first measured this force toward the end of the eighteenth century using a similar device to one that enabled him to measure the force of friction between two bodies.

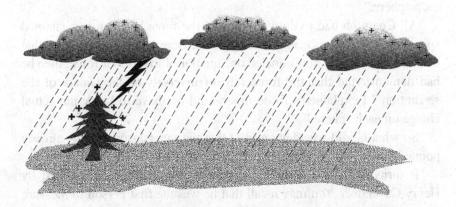

Fig. 9. The negative charges (–) at the bottoms of clouds repel electrons in the ground below, rendering it electrically positive (+). Using the wet air as a conduit, electrons forming an electric discharge then flow from a cloud to the wet tree on earth.

He demonstrated that the electrostatic force was inversely proportional to the square of the distance separating the two charges and directly proportional to the product of their magnitudes. You may note that the mathematical expression for the force between two charged bodies is the same in appearance to that expressing their mutual force of gravitational attraction. The electrostatic or Coulomb force is so much stronger than the gravitational one, however, that it would take at least 34 zeros following a one to express their ratio!"

"How did Coulomb know what the actual charge on a body was?" Beth inquired. "Who had measured this charge?"

"Let's come back to this later. We now know that the basic unit of electric charge is that of one electron, which was not discovered until the end of the nineteenth century. So Coulomb had to be extremely inventive to establish the force relationship now bearing his name. What he did was to place identical metal spheres in touch with each other and then charged them by contact with an electrically charged metal pin. Assuming the charges transmitted from the pin were equally divided between the spheres, he could establish that the charge on each sphere was 1/2, 1/3, 1/4, etc., of the initial charge. By varying the distance between pairs of these charged spheres and measuring the forces that they exerted on each other, he could verify the correctness of his mathematical expression."

"I still don't see how he did that without knowing the actual charges on each sphere."

"All Coulomb had to do was measure the force between two charged spheres at one distance and repeat the measurement at, say, twice that distance. If the force now equaled one quarter of the former force, then he had demonstrated that the force fell off inversely as the square of the separation." I explained "To do this he did not have to know the actual charge on each sphere."

"So what he did was demonstrate that the inverse-square law, already pointed out by Newton, also held for electric charges."

"It turns out that Coulomb's discovery was actually anticipated by Henry Cavendish. You may recall that he was the first person to measure the 'weight' of the earth. An inheritor of considerable wealth, Cavendish lived in relative isolation and published very few papers. So it was not until many years later that his unpublished manuscripts revealed this pioneering work. That this discovery would be made was inevitable. The world has rightly credited it to the one who published it first."

"Did Cavendish succeed in discovering the unit of charge you called an electron?"

"No. Electrons were not identified until the experiments carried out by another Englishman, J. J. Thomson, about a century later. Using a cathode-ray tube, which is the forerunner of the modern-day television tube, Thomson established that the relative charges borne by electrically charged gas atoms in his tube were multiples of the charges of the cathode rays themselves. He proposed the term *electron* for this unit of charge and demonstrated that this had to be the smallest unit of electric charge possible even on an atomic scale. The limitations of his apparatus, however, enabled Thomson to measure only the ratio between the charge and the mass of an electron but not their respective values. This was done by a University of Chicago professor in 1913."

"Since the two mathematical expressions for the force have the same form," Beth interjected, "is the Coulomb force similar to the gravitational force described by Newton?"

"No, they are actually quite different. In fact, the electric force is so much larger than the gravitational one that the latter can be totally ignored by comparison. More importantly, the Coulomb force can be either attractive or repulsive whereas gravitation is an attractive force only."

"Well," Beth persisted, "is there a similar unit of gravitational force, analogous to the electron?"

"Many other people have asked the same question. Rather elaborate experiments have, in fact, been under way for some years in the hope of catching such a *graviton*; without any success to date, I might add. But let's stick to the discussion of electric charges in the eighteenth century and see how we made use of them."

## Storing Electricity

"Some of the electrostatic generators built in the eighteenth century could accumulate sizable charges so that, upon discharge, they produced spectacular but very brief electric sparks." I continued. "It wasn't until the end of that century, however, that the ability to sustain a continuous flow of electricity, called an *electric current*, was developed. It's of passing interest that the initial discovery, like some others during the early stages of our science, was made by someone whose formal training was not in the physical sciences at all.

Thus it happened that the laboratory established for the dissection of animals at the University of Bologna in Italy housed an electrostatic generator on the same long bench on which the anatomy lessons were pursued. By turning a crank, the circular disk in the generator rubbed against a pad connected to a metal globe in which the electric charges produced would accumulate. One day, while an assistant of anatomy professor Luigi Galvani was dissecting a dead frog, he noticed that the muscle of the frog's leg contracted when he touched the tip of a metal scalpel to the inner nerve of the leg. A student working nearby remarked that this occurred only when he happened to generate an electric spark in the electrostatic generator."

"Was the electric charge transmitted through the air, the table, or what?" Beth wanted to know.

"Actually the transmission took place through the air but a hundred years had to pass before physicists discovered how that can happen. After this strange observation was reported to Professor Galvani," I went on, "he proceeded to explore their finding in the systematic way characteristic of scientists. First he duplicated his assistant's actions and observed the same contractions whenever the electric discharge coincided with the contact of a metal to the frog's nerve. Next he repeated the tests by using different metals, with and without accompanying electric sparks. In this way Galvani discovered that the muscular contractions could be produced by touching the leg by two different metals that were joined together at their opposite ends. If the frog's leg was contacted by two rods made of the same metal, then a contraction did not take place."

"I like the point you made about the way scientists first verify that an original observation is reproducible and then go on to see under what other circumstances the same thing might occur." Beth mused. "We do the same thing, of course, in psychology as well."

"In pursuing his studies," I resumed, "Galvani tried many other ways to stimulate the muscular contractions in the dead frogs, including suspending them from brass hooks attached to an iron fence in his garden. During a lightning storm, he observed contractions taking place whenever lightning was seen somewhere in the sky. On a dry day, Galvani found that he could produce a similar contraction by touching a frog's leg to the iron fence. After seemingly exhausting the various possibilities, Galvani published his carefully documented observations in 1791. In this document he attributed the observed contractions to a 'force of animal electricity' that must be inherent in the frog."

"Surely that's not the correct explanation of Galvani's observations?" Beth asked.

"No it's not. But, as sometimes happens in science, a naive early explanation turns out to have a germ of truth in it that is not appreciated until much later."

"Oh, I bet you are referring to some of the more recent discoveries by neurobiologists that neurons in living bodies transmit electric impulses. Aren't you?"

"Right again! This particular discussion reminds me of the story about the scientist who, while attending a conference in Paris, also visited a flea market there. After purchasing a 'trained' flea in the market, he brought it to his room and set it down on a table. Using a stop watch, he then determined that, each time he ordered the trained flea to jump, it would rise about three inches in the air. Next he cut off one of the flea's rear legs. Now when told to jump, the flea could only rise about half inch off the table. Finally he removed the flea's other leg and ordered the flea to jump. The poor flea now could not lift its body at all. Recording all this dutifully in his notebook, the scientist then added this conclusion: 'when both hind legs are removed from a flea — it loses its hearing'."

"That's a good story," Beth said, while laughing at the joke. "It certainly illustrates how more than one explanation of the same observation can appear to be plausible."

"That's right. Although the correct explanation for the mythical flea's inability to jump is pretty evident, Galvani's explanation of his observations seemed quite appropriate to one trained in medicine in the eighteenth century. These observations reached the attention of a contemporary professor of physical science," I returned to my tale, "who reinterpreted Galvani's observations correctly in terms of moving electric charges. Alessandro Volta, who was investigating what happened when two different metals were placed in contact with a liquid, naturally was able to attribute the contractions to the fact that the metals touching the frog were dissimilar rather than to any inherent properties of frogs. Not only was this correct explanation confirmed by further developments, but it enabled Volta to extend his observations to applications beyond twitching frog legs. His researches, by the way, became so well known throughout Europe, that they even elicited an invitation to visit Napoleon in Paris. There the emperor bestowed upon Volta a special medal commemorating this occasion."

"Is this why we measure electricity in volts?" Beth wanted to know.

"Yes, and in aminute, I'll tell you all about electric terms." I responded. "First I want to finish my tale about Professor Volta."

"In a letter sent to the President of the Royal Society of London in 1800, Volta described a new 'apparatus..., which will doubtless astonish you, is only an assemblage of a number of good conductors of different sorts arranged in a different way.' This apparatus, Volta went on, could produce sparks having much greater intensity and noise and could jump across much larger gaps than those produced by any other devices. More importantly, the new one differed from others in that 'it does not need, as they do, to be charged in advance by means of an outside source; and in that it can give the disturbance every time it is properly touched, no matter how often'."

"What Volta was describing," I continued, "was a precursor of the modern electric battery. It was composed of individual cells, with each containing two dissimilar metals in contact with a liquid called an electrolyte. Modern day batteries differ only in the kinds of metals and liquids that they contain. Even the so-called 'dry' cells used to power a flashlight or a portable radio contain an electrolyte which can leak out of its container."

"Is such leakage why manufacturers advise the removal of dry cells from appliances not in frequent use?" Beth interrupted.

"Yes, that is exactly why."

"What's the difference between a cell and a battery?" Beth now wanted to know.

"A cell consists of two metals or electrodes immersed in a liquid electrolyte, one of which becomes electrically positive while the other is

Fig. 10. Volta's first battery consisted of glass jars filled with an electrolyte into which copper (A) and zinc (Z) rods were immersed. The unlike metal rods in adjacent jars (cells) were linked to each other as shown in his original drawing.

negative. A battery is made up of a series of cells joined to each other so that the positive electrode of one cell is connected to the negative electrode of the next cell (Fig. 10). When this is done, the voltage generated in one cell adds directly to the voltage generated in the adjacent cell so that the total voltage of the battery increases with the number of cells that it contains."

"I realize that it is named after Volta, but what exactly is this 'voltage' of a battery?" Beth now inquired.

"What happens in each cell is that a chemical reaction between the two metals and the liquid electrolyte causes positive charges to accumulate on one of the electrodes and negative charges on the other. We now realize that these accumulated charges are stored in the cell until they are discharged by enabling negatively charged electrons to flow from the negative electrode to the positive one. The flowing electrons make up an electric current which, as you know, can do work by converting its electrical energy to heat or light or some other form of energy. Thus the battery provides a way of storing electrical energy so that this electrical energy is a form of potential energy."

"Do you remember from our earlier discussions that energy is nothing more than a mathematical construct?" I continued. "This means that now we can take the potential energy accumulated at each electrode and divide it by the magnitude of the unit charge of one electron and call this new quantity the *electric potential* that a charge 'feels' in the vicinity of that electrode. When an electron flows from the negative to the positive electrode, it undergoes a change in its electrical potential by an amount called the *potential difference* between the two electrodes. In memory of its discoverer, this quantity is measured in volts. Thus the potential difference and the voltage of a battery are merely two terms for the same thing."

"So, the voltage of a battery is a physicist's way of relating its electric potential energy to one electric charge," Beth concluded. "How does this translate into the electric current of moving electrons?"

"An electric current is measured by noting the number of charges that move past a point per second. The unit selected for this is called an *ampere*, after a French physicist who played a key role in relating magnetism to electricity."

"By the way," I went on, "the charge carriers can be the electrically charged atoms or *ions* in a liquid electrolyte or the electrons that are also present in an electrolyte as well as in anything else that is made up of atoms."

"Electrons move in response to a potential difference expressed in volts and this produces a current measured in amperes," Beth summarized neatly. "Surely physicists must have a way of relating a volt to an ampere?"

"You're quite right and thereby hangs another interesting tale. A German high-school teacher, Georg Simon Ohm, reported in 1827 that his systematic studies had established a very simple ratio between the voltage applied to a wire and the current produced in it. When the voltage, in volts, is divided by the current, in amperes, the ratio is a constant that is characteristic of the wire or other circuit element to which this potential difference is applied. This ratio actually measures the *resistance* to the flow of electrons or current in the conductor and we now express this resistance in ohms. In the early nineteenth century, however, this elegant relationship proposed by Ohm was not considered to be reliable. Even worse, Ohm was considered to be unreliable for proposing it so that he actually retired from teaching soon thereafter."

"Why was it considered unreliable?" Beth wanted to know. "I thought that physicists loved simple explanations of natural phenomena."

"In part it was a matter of the skepticism with which we humans view anything thought to be 'too good to be true'. In part it was caused by the difficulty of measuring the resistance precisely because an electric current actually causes the resistance of the conductor to change with time.

But this story has a happy ending. It took only a few years for Ohm's formula, now known as Ohm's Law, to be proven correct. This led to his appointment as professor of physics at the Polytechnicum of Nüremburg in 1833 and a few years after that to a chair at the more prestigious University of Munich."

"I wonder whether they doubted poor Georg Ohm because he was a high-school teacher and not a member of the physics elite?" Beth commented as we rose to clear away the breakfast dishes. "But truth seems to have won out after all!"

# Chapter Eight:
# Breakfast of Blueberry Muffins

## *Electricity in Matter*

"To celebrate the return of electricity in our house yesterday, I baked us some blueberry muffins for breakfast." Beth reported the next morning. "As I was mulling over what you told me yesterday, I began to wonder: does the material the current passes through make any difference?"

"Indeed it does. Even when the same potential difference is applied to two different materials, the currents produced in them will not be the same. For example, the different resistances exhibited by what we call conductors and nonconductors are determined by their atomic structures. As we'll discuss in more detail later, atoms are made up of an electrically positive inner nucleus that is surrounded by a cloud of negative electrons. When atoms come together to form matter, it is the interaction between them that determines the nature and the properties of that matter."

"I see now why you stressed the relative strength of the electrostatic or Coulomb force of attraction," Beth exclaimed. "It's the force that binds the electrons to the atom's nucleus and one atom to another in a solid."

"Absolutely correct! Nevertheless, the actual interactions between atoms can take many different forms. In fact, this is what gives matter its various properties. For example, you know that you can hold a hot skillet by its nonmetallic handle without getting burned. That's because the handle is a nonconductor of heat while the metal is a very good conductor."

"So, the metal in the skillet is a conductor of heat while the plastic handle is a nonconductor?" Beth asked to confirm her new understanding. "Does the same apply to conductors and nonconductors of electricity?"

"Yes." I replied. "As we shall see, the same electrons that conduct the one are also responsible for conducting the other. Let's consider first what happens in a good conductor like a metal. The atoms in metals are actually positive ions made up of an inner nucleus surrounded by a cloud of electrons whose total negative charge is somewhat less than the positive charge on the nucleus. This is so because the electrons that would render individual atoms electrically neutral have been removed from individual atoms and are shared by the metal as whole. In this way, these electrons can be thought to form an *electron gas* that permeates the entire metal (Fig. 11). The Coulomb attraction between the positively charged atoms or ions and the negative electron gas is what holds this entire assemblage together. At the same time, the relative freedom of the electrons making up the electron gas accounts for all the distinguishing characteristics of metals. One of these, as you well know, is that metals are excellent conductors of electricity."

"Is that because the electrons making up the electron gas are free to move in response to an applied potential difference?" Beth asked.

"Exactly! By the start of the present century, a *free-electron theory* was formally proposed to explain electrical and thermal conductivity in metals as well as most of their other properties, which we can talk about some

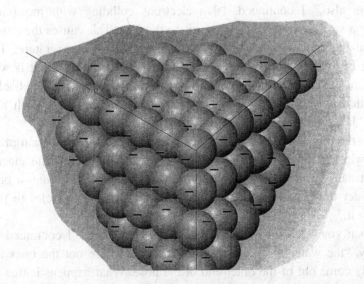

Fig. 11.   A metal is composed of a highly regular array of positively charged metal ions (shown as spheres) surrounded by a gas made up of negatively charged electrons.

other day. According to this theory, the electrons making up the gas are constantly moving about so that there are actually small electric currents flowing in the metal at all times. Because their motions are random, however, there are as many electrons moving in one direction, on average, as there are moving in opposite direction. In this way, no net-current results. When a potential difference is applied to the metal, all of the free electrons tend to flow toward the positive end. As they do this, however, individual electrons collide with the metal ions which scatter the electrons and this impedes their forward motion."

"And that's what causes the metal to have a resistance," Beth interposed.

"I'm delighted by how quickly you're catching on. I should have pointed out that the ions are actually not fixed in space but vibrate about their ideal locations by small amounts dependent on the temperature of the metal. The higher the temperature of the metal, the larger does the amplitude of these vibrations become and the more likely, therefore, any collisions with the moving electrons. This, of couse, increases the metal's resistance to electron flow so that the electric current in a metal declines as its temperature is raised."

"Note also," I continued, "that electrons colliding with metal ions transfer a small portion of their energy to the ions. This causes the ions to increase their vibrations, thereby raising the metal's temperature. This heating of a conductor by the electric current passing through it is what caused the deviations from Ohm's law that made it appear to be unreliable."

"Is that why electric wires sometimes get so very hot?" Beth now asked. "Is there any way to ease this heating?"

"Yes there is. All we have to do is to increase the cross section of the metal wire. In this sense, the passage of electrons through an electric conduit can be likened to the flow of water molecules through a pipe. The wider the conduit, the easier it is for the moving particles to pass through it."

"When you open the spigot at one end of the hose," I continued the analogy, "the water molecules that enter the hose are not the ones that instantly come out of the other end of the hose. What happens is that the momentum of the entering molecules is passed on to the neighboring molecules that, in turn, pass it on to their neighbors, and so forth down the length of the hose. The water molecules at the opposite or outlet end then get literally pushed out of the hose and we say that the water is flowing through it. Obviously, the larger the cross section of the hose, the more water molecules can pass through it but their speed is not affected by this."

"When I turn on a water faucet in the garden," Beth observed, "it takes the water a short time to flow out of the other end. When I flip a light switch to *ON*, however, the lamp connected to it lights instantly. Obviously, electrons must move much, much faster than water molecules, or is there some other explanation?"

"Your analysis is quite correct. The reason there was a time delay before the water ran out of a hose also was because it was not entirely filled with water at the moment when you turned on the faucet. What happens in a full pipe or hose can be illustrated by one of my favorite toys consisting of a row of metal spheres suspended from a bar by strings of equal length (Fig. 12). If one moves the right end ball to one side and lets

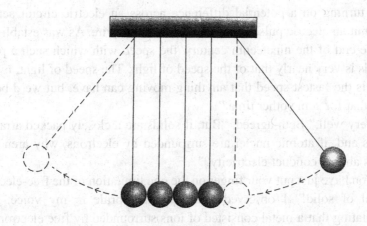

Fig. 12.   If the end ball of the suspended row of touching balls is moved off to the right and then allowed to fall back, the ball at the opposite end will be caused to fly off to an equal distance to the left in order to conserve the total momentum. What will happen if three balls are removed and then allowed to fall back together? (Answer: three balls must fly off from the opposite side.)

it go, the ball at the opposite end will fly off at the instant when the right ball returns and strikes the other balls in the row. The left end ball then flies off and, upon its return, the right end ball flies off again, and this process repeats itself until frictional losses gradually dampen the process down. Can you tell me what is causing this?"

"Isn't this a case of momentum conservation?" Beth asked. "I seem to remember that removing two balls at one end then causes two balls to fly off at the opposite end, and so forth."

"Right again Beth. You really ought to give up psychology and take up physics."

"No thank you. One of the reasons why I find the analysis of human behavior so challenging is precisely because it is not possible to predict it with the same assurance that makes you physicists feel so smug. But," Beth persisted, "you still haven't explained why a light goes on the instant I switch the electricity on."

"What happens when the end ball falls back and strikes the above row of suspended balls? We think of it as a transmission of its momentum through the balls in the row to the ball at the opposite end. In the same way, turning on a potential difference across an electric circuit acts to transmit an electric pulse down the length of the wire. As was established by the end of the nineteenth century, the speed with which such a pulse travels is very nearly that of the speed of light. The speed of light, by the way, is the fastest speed that anything moving can have, but we'd better save that for some other time."

"Very well," Beth agreed. "But, if solids are a closely packed array of atoms and, if atomic nuclei are surrounded by electrons, why aren't all solids able to conduct electricity?"

"You have just put your finger on the one limitation of the free-electron model of solid!" I observed with a certain pride in my voice. "By postulating that a metal consisted of ions surrounded by free electrons, it had no way of explaining why all solids weren't metals. Obviously, a different model had to be invented and it gradually evolved during the late 1920s. To begin, we already know from Coulomb's law that the electrostatic attraction between the positive nuclei and negative electrons surrounding them varies with their separation. As a result, the energies that the different electrons in atoms have are not all the same.

Furthermore, when atoms join to form solids, the additional forces exerted on their outer electrons by the surrounding atoms modify the relative energies that these electrons can have still further."

"Let me be sure that I am following what you are saying," Beth interrupted. "The electrons surrounding a positive nucleus of an atom do not all have the same energy. Right? Then, when the atoms are packed together in a solid, these electron energies are further changed because there are now neighboring atoms that also exert forces on them."

"Well said!" I virtually beamed. "As we have seen, in metals, the shared electrons making up the free-electron gas can assume a continuous range of possible energies. These electrons are free, therefore, to gain kinetic energy from an external electric potential and to move in response to it. In all other materials that are not metals, the electrons of each atom remain more tightly bound to their parent atoms. In nonmetallic solids, this limits the energies that electrons may have to discontinuous sets of values. These can be represented pictorially by drawing parallel bands of allowed energies separated by bands of forbidden energies (Fig. 13)."

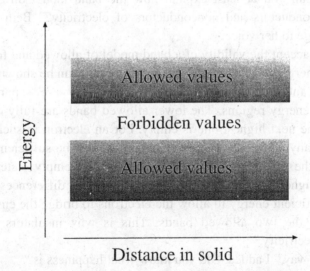

Fig. 13. The energies that electrons can have in solids can be represented by energy bands extending throughout the solid. The shaded regions, representing allowed energies, are separated by regions containing energies that electrons cannot have.

"I'm not sure I understand this picture," Beth interjected. "What you are calling 'allowed values' are the actual energies that electrons in that solid can have. What then prohibits them from having the other energy values you have labelled 'forbidden'?"

"Let me hold off answering this question until we have had a chance to discover more about the structures of atoms. For present purposes, please accept the fact that such a *band model of solids* is a useful way of explaining the physical properties of solids. I should alert you, though, that the existence of allowed and forbidden energies was not conceivable until the invention of the quantum theory at the start of the present century. This means that, during the first-hundred years that followed Volta's original creation of a continuous electric current, no one really knew what an electric current was. This relative ignorance did not prevent, however, the discovery of numerous practical applications of electricity. Once a complete understanding of electric phenomena was established in the present century, of course, a truly revolutionary set of developments was made possible. This, as you know, has completely changed the lifestyles of people all over the world."

"Well, can you at least explain how the band model differentiates between conductors and nonconductors of electricity?" Beth inquired with an edge to her voice.

"If you accept the validity of a band model of allowed and forbidden electron energies," I hastened to respond, "then it can be shown that the allowed bands in nonconductors of electricity are separated by forbidden energy regions. The lower allowed bands are fully occupied whereas the next highest one is empty. For an electron in such a solid to absorb any additional energy, this energy must be sufficiently large to allow the electron to move into one of the empty states in the allowed higher energy band. Ordinary potential differences do not supply sufficient energy to allow the electrons to bridge the energy gap separating the two allowed bands. This is why insulators can not conduct electricity."

"By the way," I added, "do you know what happiness is?"

"What's your definition?"

"Having electricity available for you to bake these yummy blueberry muffins!"

"I'm enjoying them myself." Beth responded. "But you haven't told me yet how the band model deals with metal conductors."

"It turns out that the band model of a metal also segregates the energies that electrons can have into bands of allowed and forbidden energy values. In a metal, however, the allowed bands overlap each other so that the electrons have available to them a continuous range of allowed energies just as they did in the free-electron model."

"That's really neat." Beth exclaimed. "The more sophisticated band model can distinguish metals from insulators without invalidating the earlier picture of why metals are conductors."

"You may not realize this, but you have just stated something that became a guiding principle in the development of modern physics in the twentieth century."

"I did?" Beth respoded with surprise. "Tell me what I said."

"As we shall see when we talk about atoms in more detail, Niels Bohr, the inventor of the quantum theory of atoms, insisted that any new theory must include in it any previous theory that correctly explained nature. This has become a guiding principle in the acceptance of some of the very radical theories developed in the present century."

"All this sounds pretty myterious to me." Beth observed. "But I must admit that my curiosity has been aroused and I can't wait to learn more about the quantum theory that, hopefully, may make this strange behavior of electrons become more comprehensible. But before we go on to that, I have some other questions about electricity."

## a.c./d.c.

"For example, what makes electricity dangerous?" Beth went on. "One frequently encounters signs warning of high voltages. I also know that I have to be careful around a car battery of only twelve volts. What exactly is the role of the voltage?"

"Good question!" I responded. "It is not the voltage but the product of the voltage and the current capacity that really matters. We call this product electric power and, as you know, we measure this power in *watts.*"

"What is a watt?" Beth couldn't help giggling.

"A watt is, simply, the product of one volt multiplied by one ampere. Because this is a relatively small quantity, the electric company reckons power in thousands of watts called *kilowatts*."

"How does this explain why a 12-volt battery in a car is dangerous but a 12-volt battery in our burgalr alarm is not?" Beth wanted to know.

"In actual applications it is important to consider the total power that a source of electricity can deliver or a device using electricity can consume." I explained. "One of the hidden assets of the solid-state electronics used in our burglar-alarm system is that they draw very small electric currents. Most solid-state electronics draw currents measured in multiples of one-millionth of an ampere, called a *microamp*, or one-thousandth of an ampere called a *milliamp*. Thus a 9-volt dry cell powering a portable radio delivers a tiny fraction of a watt to the radio. By comparison, a starter motor in an automobile may require as much as ten watts of power, so that a 12-volt car battery has to be able to deliver current of at least one ampere. That is why you are quite safe when handling the small batteries you buy in the drug store, but must exercise caution around a car battery."

"What decides how much power is actually drawn from a battery?" Beth asked in a doubtful tone.

"The total resistance of the electric circuit connected to it." I responded. "Should you accidentally touch the positive terminal of a car battery while wearing wet shoes, you will provide a relatively low-resistance path for the electric current. In accord with the empirical law of Georg Ohm, a low resistance draws a relatively large current. Even the one-hundredth of an ampere current generated in a human by a 12-volt car battery could cause an involuntary heart spasm and that can be deadly."

"Is that the same reason why we see warnings against handling electric appliances while in the bath tub?" Beth asked.

"Absolutely!" I replied. "Again, the danger comes from appliances plugged into the house circuit which can deliver kilowatts of power. A battery-operated radio in a shower is quite safe because it can only deliver microwatts of power."

"Isn't it true that electric fuses or circuit breakers protect us from sudden current surges?" Beth asked next.

"A fuse is a device containing a very thin wire which can transmit safely only a finite current. When this current rating is exceeded, the wire gets so hot that it melts and interrupts the flow of electricity. A circuit breaker accomplishes the same purpose, but it doesn't destroy itself. While fuses or circuit breakers will prevent excessive currents from damaging sensitive equipment, they will not save a human from electrocution by a current smaller than that required to activate a fuse or breaker. By the way, before resetting such a circuit breaker, it is important to remove the cause of the current surge that caused the breaker to trip in the first place."

"I'm not clear about one thing," Beth interjected. "To produce a potential difference and, hence, an electric current, it is necessary to connect to both the positive and negative terminals of a battery. How then does a person touching only the positive terminal of a car battery run risk of electrocution?"

"The negative terminal of a battery is electrically connected to the car's body. A wet car provides a direct path for an electric current to the ground on which the car and its operator usually stand. Thus, by touching the positive terminal or anode, one provides a continuous circuit from the anode, through the body to the ground, back to the negative terminal."

"In order to produce a current," I went on, "a potential difference must exist. This means that a bird can sit safely on a high-tension wire as long as both its feet are touching the same wire. In this case both feet are at the same potential. A squirrel, attracted to the high-tension wires by the hum or warmth of a transformer mounted on a pole, may straddle two of the lines between which there exists a considerable potential difference. At that instant, the poor squirrel becomes electrocuted."

"What about electric circuits in our house?" Beth persisted. "Are they connected to the ground? And, if so, aren't they dangerous?"

"As a matter of fact, they are connected to the ground but this makes them safer."

"How?"

"The negative side of the electric power line entering our house is directly connected to the earth through a special 'grounding' rod or by being connected to a metal water pipe that goes into the ground." I continued. "Any current surge appearing in a house circuit thus is directed to the ground, where it is dissipated harmlessly. That is why power tools, hair driers, laundry irons, television sets, and other devices drawing large currents include separate wires that also go to the ground. Called grounding wires, they prevent a possible potential difference from arising between the appliance and the house circuit, which would endanger an unwary user of the appliance."

"Is that why such appliances have electric plugs that only fit one way into the wall outlets?"

"Yes," I replied. "One side of the plug always goes to the wire that is directly connected to the ground. It is very important not to circumvent these safeguards because they are there to protect you from possible dangers."

"Safeguards like that are really impressive." Beth observed. "But who thought up all the gadgets that we now take for granted?"

"Much of the credit belongs to Thomas Alva Edison."

"He really was amazingly accomplished, wasn't he? I remember learning about Edison in school and I really enjoyed our visit to the Edison Museum in Ft. Myers, Florida." Beth chimed in. "I remember that Edison quit school at the age of twelve and ran off to work on a railroad. He became a telegraph operator at the age of fifteen and took out his first patent at the age of twenty one. By the time he was eighty, Edison had accumulated over 1,000 patents, which must be some kind of a record!"

"Yes, he was quite a guy!" I observed. "Although some of my colleagues tend to sneer at his unscientific methods and his lack of formal education, there is no way to minimize the importance of his manifold contributions."

"Probably because he used a trial-and-error approach instead of attempting to develop an elegant theory first, a lot of 'true' scientists have branded his work as unscientific." Beth hinted. "Could it be that such critics are just a wee bit envious?"

"More than likely." I agreed.

"Edison must have gotten some more schooling along the way," Beth insisted, "in order to have made all those inventions."

"As far as formal schooling goes, Edison had none after he quit his school in Michigan. He certainly kept up with scientific developments of his time but, later in life, he would boast that he didn't need to know any science because he could hire scientists any time he needed them. Edison maintained that the first step in the invention process was to recognize a human need for some device. The rest was imagination and perseverance."

"Sounds easy when you say it that way," Beth observed.

"The hard part is the perseverance," I responded. "Take the electric light bulb. Edison did not actually invent it, although a lot of people think he did. He knew about the earlier experiments with incandescent lamps that had taken place in the nineteenth century, chiefly but not exclusively in England. Most of those bulbs worked all right but their filaments burned out so quickly that they were not practical. So Edison's challenge was to find a long-lasting filament. By the time he had made a successful light bulb utilizing a carbonized cotton thread, Edison had tested over one thousand different materials, including a red-headed Scotman's hair! So, you see, there was indeed nothing scientific about what my sneering colleagues today refer to as 'Edisonian research'."

. "But, to be fair, maybe Edison short changed himself by not knowing more about physics." Beth opined. "Had he understood better how electricity works, he might have avoided many of his false starts. Surely, this must have been pretty costly."

"Yes it was. Edison spent many thousands of his own dollars before he came up with the right bulb design, a truly princely sum in 1879. He spent even more of his own money during the next ten years, when he not only continued improving his light bulb, but also developed the means of generating, storing, and distributing electricity. And, most importantly, Edison patented each and every step along this process."

"So that's why so many electric power companies have incorporated Edison's name in their own," Beth commented.

"That is quite true and ironic at the same time. I say ironic because Edison was a strong believer in what we now distinguish as *direct* current, like that produced by batteries. Yet nowadays, most electric utilities transmit alternating currents."

"What's the difference?" Beth immediately inquired.

"A direct current, abbreviated *d.c.*, consists of electrons that flow from the negative electrode, or *cathode*, to the positive electrode, called an *anode*. Hence the name *direct* or one-directional current. An alternating current, or *a.c.*, is produced when the electric potential changes back and forth or *alternates* so that the electrons first flow in one direction and then have to reverse their direction as the applied potential changes its direction. In the United States, by the way, an altenating current changes its direction 60 times each second."

"Why do we prefer to use *a.c.* to *d.c.*?" Beth naturally asked at this point.

"The reason for our preference of *a.c.* has to do with the magnetism that accompanies all electric currents. We haven't had a chance to discuss magnetism yet but we can take that up at our next breakfast. Besides, if I eat any more muffins, I may have difficulties getting up from the table."

# Chapter Nine:
# Breakfast of Apple Fritters and Love

## Magnetic Forces

"I believe that you are going to tell me how magnets make love?" Beth teased the next morning between servings of orange juice and apple fritters drenched in real maple syrup.

"Oh, you must be thinking about Dr. Gilbert's term: magnetic coition." I reacted.

"Is that where your mind is this morning?" I couldn't resist adding. "While Eve tempted Adam with an apple, are you going her one better by tempting me with these delicious apple fritters?"

"I wonder what Gilbert had on his mind'?" Beth asked, bringing the subject back to her original question. "Please tell me more about magnetism."

"The history of magnetism goes back about twenty-five-hundred years when, according to a Greek legend, a shepherd in a region called Magnesia noted that the iron crook on his staff was strangely attracted to an outcropping of lodestone, which is the common name given to a mineral form of iron oxide. As a result, lodestones exhibiting such properties subsequently were named *magnets*. Romans learned about them centuries later but, surprisingly, there are no records indicating that anyone recognized the ability of magnets to align in a north–south direction. The first recorded use of this property of magnets comes from China in the eleventh century A.D. It is most likely that Muslim traders sailing to China actually were the first to use a compass to help them navigate. Two more centuries had to pass before European navigators learned this art. Soon sailors became so dependent on the compass that tampering with one on shipboard was deemed a crime as serious as mutiny."

"That's all very interesting but can we get back to love making by magnets?" Beth nudged me whimsically.

"Dr. Gilbert was the first to set out the distinctions between the behavior of magnetic lodestones and amber, which is capable of being charged electrically. You may recall that one can charge some substances so that they become electrically positive and others electrically negative, but never both. A piece of magnetic lodestone, in distinction, has opposite ends that behave differently. If freely suspended, one end will point toward the earth's north pole and the other end will point toward the south pole. If you break the magnet in half, each half will have a north- and south-seeking pole, and so forth, no matter how small you make each magnetic piece."

"What happen when two magnets meet?" Beth persisted.

"Gilbert showed that the north-seeking end or *pole* of one magnet attracts the south-seeking pole of another magnet whereas two north or two south poles repel each other."

"Well, if a magnetic compass needle always points toward the earth's north pole, that implies that the earth itself is a magnet and that it is the south pole of the magnet that points to the north pole of the earth." Beth concluded triumphantly.

"Good thinking!" I responded with pleasure. "Actually Gilbert's discovery in the sixteenth century provided a rational basis for future actions by navigators, explorers, and map makers alike."

"And after that?"

"About a hundred years later," I responded, "Coulomb established with considerable precision that magnetic attraction and repulsion obeyed an inverse-square law like the one that applied to electrostatic forces."

"I see now why you stressed the significance of Newton's original postulate of the inverse-square law in his theory of gravitational attraction." Beth interrupted.

"That's right." I continued. "It is helpful in discussing magnetism to think of the magnetic forces acting along lines that extend from the north to the south pole of the magnet (Fig. 14). There are at least two ways that

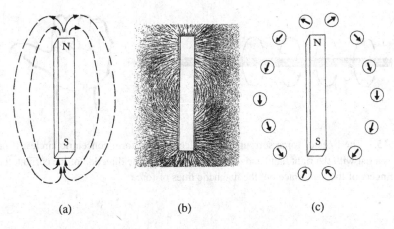

(a)                    (b)                    (c)

Fig. 14.   (a) The lines of magnetic force passing from the north to the south pole of a bar magnet actually continue inside the bar to form continuous magnetic loops. (b) Iron filings surrounding a bar magnet align themselves along the magnetic lines of force. (c) Compasses placed around the bar magnet indicate the direction of the magnetic force lines.

one can actually visualize these lines of force. One is to place a thin piece of paper over the magnet and then sprinkle iron filings onto the paper. When this is done, the iron particles array themselves along the lines of force rendering them readily visible (Fig. 14(b)). Alternatively, one can place small compasses around the magnet in which case their magnetic needles align along the lines of force also."

"That's neat!" Beth exclaimed. "I can actually see the forces stretching from the north to the south pole of a magnet!"

"About thirty-five years after Coulomb announced his force law and some twenty years after Volta described his electric battery," I resumed my narrative, "a momentous event took place in a lecture hall of the University of Copenhagen in Denmark. Hans Christian Oersted was in the midst of a lecture-demonstration of magnetic properties when he noticed that his compass needle was deflected whenever he switched on the current in a nearby electric conductor. It would have been quite natural for him to brush off such a distraction from his carefully prepared lecture; but for quite a while Oersted had been seeking unsuccessfully a relationship between electricity and magnetism.

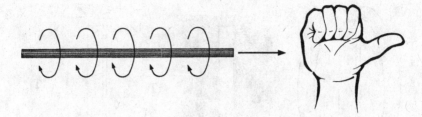

Fig. 15.  The lines of force surrounding an electric wire form concentric circles. If one forms a fist with the right hand and extends the thumb in the direction of the current, then the fingers of the hand trace out the magnetic lines of force.

Excited by this chance observation, Oersted got hold of a more powerful battery and devoted himself to trying to understand his unexpected observation. By noting the deflections of a compass needle at different points around a wire carrying an electric current, he established that magnetic lines of force form rings around the wire (Fig. 15). Then Oersted invited a number of distinguished countrymen, including the King's Minister of Justice, to his laboratory to witness this phenomenon. They all were visibly impressed."

"Am I correct that Oersted also was a distinguished poet in his day?" Beth asked.

"Yes, but he is far better known as the first to demonstrate the intimate relationship between magnetism and electricity. In fact, the American Association of Physics Teachers awards an annual Oersted Medal to commemorate the momentous discovery made in the midst of a lecture on physics."

"I also seem to remember that Benjamin Franklin had tinkered with magnets in his day." Beth contributed. "Didn't he report that iron needles become magnetic when in the presence of a nearby electric discharge? In fact, in his novel *Moby Dick*, Herman Melville described a dramatic episode in which lightning reversed the magnetization direction of a compass needle."

"You are a living demonstration, my love, that the arts and sciences do not have to exist in two separate worlds!" I observed most proudly.

"You betcha! But let's return to Oersted."

"Sure. I can tell you that Oersted's discovery triggered a great deal of interest in scientific circles around the world. In Paris, the French mathematician and theoretical physicist, André Marie Ampère, demonstrated that two parallel current-carrying wires attract each other if they carry currents in the same direction and repel each other, if their currents move in opposite ways. I might add that Ampère was a typical absent minded professor. Not only did he fail to recognize the Emperor Napoleon Bonoparte when he visited his laboratory one day, but Ampère then forgot to show up at the emperor's palace for a dinner party the following evening!"

"Maybe they had similar magnetic poles?" Beth tried to keep a straight face. "Going back to the magnetic lines of force, do the tiny iron filings line up near a magnet because they too have become magnetized?"

"Exactly! Several substances can become magnetized when placed in a magnetic field. Iron is best known for this property and, because it is plentiful and inexpensive, iron is most widely used in magnetic applications.

What happens in iron and other like materials is that the iron atoms, acting like tiny magnets, align parallel to the lines of force. Because of properties unique to iron, these tiny magnets remain aligned even after the external magnet is removed. In fact, this phenomenon is named *ferromagnetic induction* after *ferrum*, the Latin name for iron."

## Magnetic Fields

"As I look at the iron filings clustered around a bar magnet (Fig. 14)," Beth observed, "I can see that they are more densely stacked near the two ends of the magnet and less near its middle and further away from the magnet."

"What you are seeing," I explained, "is the decline of the magnetic force with distance from either end or pole of the magnet. This can also be described as a magnetic field that surrounds the magnet. Its strength falls off inversely with the square of the distance to the poles."

"This is just like the gravitational field of Newton or the electrostatic field discovered by Coulomb." Beth concluded. "The lines of force that we talked about before are then nothing but imaginary traces of such a field, right?"

"Quite right. Keep in mind, however, that a gravitational field surrounds any concentration of mass, like the earth, and its lines of force are attractive only to some other mass. On the other hand, an electric charge is surrounded by an electric field that can be positive or negative so that it may be either attractive or repulsive to another charge. Finally, a magnetic field is a *dipolar* field, so named because it always surrounds both a north and a south pole of any magnet. The interaction of such a field with another magnet, consequently, will depend critically on how and where the other magnet is placed."

"This is becoming a little confusing," Beth admitted, "although I can see the general idea."

"I know. The physics of magnets and, particularly, the interplay between electric currents and the magnetic fields that surround them is both conceptually and mathematically complex. On the other hand, it goes to the very heart of our understanding of physics and all the marvelous developments that subsequently followed. Since I had a great deal of difficulty with this subject while a student myself, I shall try to keep our discussion just as simple as possible."

"I wonder if Professor Oersted realized what he was on to when he proudly showed off his discovery in Copenhagen?" Beth wondered.

"Following this discovery, scientists in many parts of the world started searching for the reverse effect, wherein an electric current would be generated by a magnetic field. One such investigator was the American school teacher Joseph Henry and another was the Laboratory Director of the Royal Institution in London, Michael Faraday. At first, the two men were totally unaware of the other's parallel efforts and so it happened that Faraday was the first to publish his results in 1831 and received official credit for them. Henry's achievements did not remain unrecognized, however. He was appointed Professor of Natural Philosophy at the College of New Jersey at Princeton in 1832 and, in 1846, Secretary of the Smithsonian Institution in Washington, D.C."

"Oh I love that museum!" Beth exclaimed. "It goes on and on with one more interesting display after another."

"So do I. We owe Henry a debt of gratitude because a lot of what we see there is a direct legacy of his early efforts to make the Smithsonian an outstanding museum and also a seat of scientific study. If we had time, I'd love to go on to describe how Henry was forced to find a place for his experiments in the basement of his small house and bootleg time from his teaching duties during the early years of his explorations. But I want to tell you more about Faraday, whose discoveries were more extensive and far reaching.

Born to an indigent family, Faraday was made an apprentice to a bookbinder at a very young age. That gave him an opportunity to learn how to read. Soon he literally devoured the various texts that he was binding. Next, he started testing some of the experiments described in the science books that he read. Soon he became sufficiently proficient to dare approach the eminent Laboratory Director at the Royal Institution, Sir Davey, and talked his way into a position as Sir Davey's assistant. There he continued his self education and made numerous and very important contributions to chemistry and physics. Ultimately, he succeeded Sir Davey and became the Laboratory Director."

"It's interesting that a self-educated, unschooled individual could do as well as a college-educated teacher of science." Beth mused aloud. "Could it be because Henry had to work in isolation in up-state New York, without the benefit of discussions with like-minded colleagues nor the physical resources existing at the Royal Institution in London? And what was their monumental discovery?"

"In one of Faraday's experiments, he wrapped two insulated copper wires into coils about the same iron ring (Fig. 16). One coil was connected to a *galvanometer*, the name given in memory of Galvani to any instrument capable of detecting an electric current. The other coil was connected through a switch to a battery. Whenever Faraday switched the current on in the first coil, the galvanometer connected to the second coil registered an electric current in it, although the two coils were not electrically connected. This current in the second coil, however, quickly died down. When Faraday turned off the current in the first coil, a current again appeared in the second coil, but this time the galvanometer registered it moving in the opposite direction!"

"Sounds like hocus-pocus to me!" Beth exclaimed.

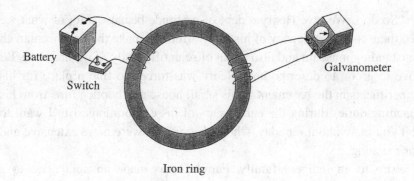

Fig. 16.   Faraday wrapped an insulated wire around a continuous ring made of iron and connected it to a battery through a switch. A separate insulated wire coil was connected to a galvanometer. Whenever the switch at the first coil was closed or opened, a brief current surge was observed in the second coil.

"Not magic at all, if you stop to examine the setup." I explained. "What happened when Faraday first closed the switch was that a direct current flowed through the first coil. The magnetic field surrounding it magnetized the iron ring which produced a magnetic field at the position of the second coil. The strength of the field went from zero to some maximum value and then remained at that value as long as the current in the first coil did not vary. When Faraday shut off that current in the first coil, this magnetic field went from its maximum magnitude back down to zero all along the iron ring."

"I get it!" Beth cried out excitedly. "It's the changing magnetic field at the second coil that induced an electric current in it!"

"That is exactly what Faraday reported in 1831: *an electric current can be induced in a conductor by the change in the magnetic field surrounding it*. The repercussions of this seemingly innocent observation have been enormous."

"Give me an example." Beth requested.

"Probably the simplest one is an electric transformer. Consider Faraday's iron ring with two insulated wire coils wrapped around it. Suppose the electric current in the first coil is *a.c.* instead of *d.c.* Every time the current changes direction in the first coil, it induces a current in the second coil even though it is not connected to a direct source of electricity. What's more, the voltage induced in the second coil is proportional to the ratio of the number of windings present in each coil. It is possible, therefore, to increase or decrease the voltage in the second coil, simply, by changing the number of windings in it."

"Oh, that's what a step-up or step-down transformer does." Beth observed with satisfaction. "And that is why electric power companies can use high-voltage transmission lines to deliver an alternating current to a local step- down transformer which changes the voltage to the 220 volts that we use here in our house."

"Right again." I observed, equally pleased. "What's more, electric transformers have an exceptionally high efficiency, with only about one percent of the power dissipated in heat losses."

"Aren't the magnetic fields of electric currents in transmission lines a possible health hazard?" Beth inquired. "A couple of years ago I read magazine and newspaper articles alleging a relationship to higher rates of cancer incidence in those living near high-power lines."

"This is a vexing question that has not been helped in the slightest by some early reactions of power companies who, of course, deny any such connection." I replied. "They contend, correctly, that the magnetic field strength is proportional to the current rather than the voltage. The current in high-voltage transmission lines is relatively low in order to minimize heat losses in these long-distance lines. Thus, the current and its accompanying magnetic field are much higher in a handheld hair dryer than in the typical transmission line.

As concluded by the National Science Academy and, even earlier, by the Connecticut Academy of Science and Engineering, analyses of the actual field data collected so far have failed to support the allegations made. A lot of the reported linkages are more anecdotal than scientifically valid. Still, they have encouraged additional researches, some of which are currently in progress. Attempts to simulate such effects under laboratory conditions have also proved negative. Nevertheless, public concern is kept high by periodic reports of new findings in the press."

"Even if a problem of some kind is established," Beth mused, "it's remedy would have to be weighed against the benefits that electric power brings into our lives. I'm sure I already know many of these, like electric motors that operate clocks, mixers, and all the many conveniences of life that we take for granted."

"Returning to Faraday," I resumed our earlier discussion, "he also invented the dynamo, which is a mass of coils placed between the poles of permanently energized magnets. When the coils are rotated by some external means, the wires of each coil cut across the lines of magnetic force. This causes a change in the magnetic field surrounding the wire and, according to Faraday's discovery, induces an electric current in each moving wire. The result of rotating the coiled wires, therefore, is that an electric current is generated in them. This makes Faraday's dynamo the precursor of modern-day electric generators."

"Is it possible to reverse the process by passing an electric current through a set of coils placed between two magnetic poles?" Beth wondered aloud.

"That is, of course, how an electric motor operates (Fig. 17). The electric current in each wire coil is accompanied by its own magnetic field. Such a field interacts with the field of the fixed magnets and this causes the coil to move or, more specifically, to rotate. And that's what makes the motor go round and round for as long as a current flows through its coils."

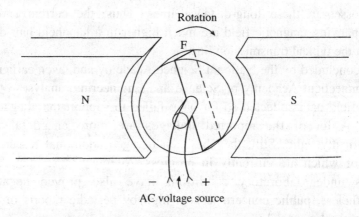

Fig. 17. An electric motor has many loops of wire, of which only one is shown. An alternating current passes through each coil. The magnetic fields thus produced by the wires interact with the steady magnetic field of the two magnetic poles producing forces that cause the wire loops to rotate about a common axis at their center.

"I am beginning to appreciate why Faraday's discoveries were so important." Beth concluded.

"Actually his interests and contributions to science were far more extensive." I observed. "Also keep in mind that Faraday was entirely self-taught. This meant that he was not as comfortable with mathematical theories as his physicist colleagues at the Royal institution. More importantly, it forced Faraday to develop mental images of how things worked and this may very well have been responsible for the many insights that his colleagues failed to get. For example, it is Faraday who actually articulated the advantages of envisaging electric and magnetic fields surrounding electric charges or magnets, respectively. In fact, Faraday generalized his conclusions to argue that the entire universe consisted of overlapping fields, including gravitational fields."

"What was the reaction of Faraday's formally educated colleagues to such pronouncements?" Beth asked.

"As you may have suspected, not favorable." I replied. "Steeped in the mechanistic concept of the universe put forth by Newton, they did not much care for what must have seemed to be a revolutionary way to view the physical world."

"People tend to be leery of new ideas. Don't they?"

"That's certainly true of most, but fortunately not of all people. And so, in the nineteenth century, Faraday's idea fell like a seed into the mind of a young and highly talented theoretician named Maxwell, where they germinated into a truely revolutionary theory whose elegance was to match that of the venerated Newton."

## What Maxwell Wrought

"By age sixteen, James Clerk Maxwell was displaying his brilliance at Edinburgh University. By the time he was nineteen, he was at Trinity College of Cambridge University, showing early promise of becoming an outstanding researcher with a keen grasp of theoretical physics. By age twenty-five, Maxwell was appointed a professor at the University of Aberdeen, where he intensified his study of electric and magnetic phenomena. Maxwell continued these pursuits after moving to a professorship at Kings College in London four years later."

"Obviously, Maxwell had a much different history from Faraday." Beth observed. "I would venture the guess that Maxwell was born in a much wealthier family."

"He was, indeed." I responded before going on. "The two actually met in 1860 and by then Maxwell already knew about Faraday's ideas about bodies exerting forces on each other over long distances.

To Faraday's experimental discovery that *changing magnetic fields induce changing electric fields*, Maxwell added the symmetrical proposition that *changing electric fields induce changing magnetic fields*. This conclusion was in direct accord with Faraday's hunch that the two fields were probably interconnected somehow."

"Did I understand you to say that Maxwell's statement that an electric field can induce a magnetic one was purely a postulate?" Beth asked.

"Although, clearly, it did not contradict any previous observations, Maxwell's postulate was just that. Keep in mind," I continued, "that Maxwell was a theoretician who analyzed and expressed physical laws mathematically.

The theory of electromagnetism that he developed boils down to just four equations (Fig. 18). Three of them are derived directly from the pre-existing laws discovered by Coulomb, Ampère, and Faraday, respectively. The fourth is the postulate cited above. Maxwell's equations involved advanced mathematics and synthesized all that had been learned about electricity and magnetism by mid-nineteenth century. In that sense, Maxwell's theory can be likened to what Newton did for mechanics two-hundred years before."

"What is the significance of these four equations that I don't understand at all?" Beth asked in a perplexed voice. "And why the comment at the bottom of the tee shirt (Fig. 18) displaying them?"

"What Maxwell's equations describe is an *electromagnetic field* that radiates outward from an electric charge moving back and forth, or *oscillating* about some point. This is a combined magnetic *and* electric field that moves outward in all directions from that point with an extraordinarily high speed. Moreover, the speed at which this field moves, according to Maxwell, turned out to be exactly equal to the best value then available for the speed of light!"

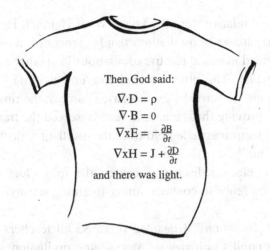

Then God said:

$$\nabla \cdot D = \rho$$
$$\nabla \cdot B = 0$$
$$\nabla \times E = -\frac{\partial B}{\partial t}$$
$$\nabla \times H = J + \frac{\partial D}{\partial t}$$

and there was light.

Fig. 18. Maxwell's equations have been used to adorn tee shirts that have become popular about one century after he first developed his theory.

"What is light?"

"When Maxwell completed his calculations, the implication was obvious: If the electromagnetic field travels with the speed of light, isn't light itself a travelling electromagnetic field? And that reminds me of a love story."

"Oh good! That's where we began."

"It seems that on the evening following this monumental discovery, Maxwell was walking with his wife-to-be. Like countless lovers before and since, she looked up at the sky and remarked on how beautiful the stars looked. Maxwell responded that he was the only man on earth who knew what starlight really was."

"Did she marry him after that?"

"Yes. She either accepted or ignored that statement. Not so his physicist colleagues. They considered Maxwell's theory far too revolutionary and, I might speculate, too difficult to understand. It took another twenty years before Heinrich Rudolf Hertz, a German professor of physics, dramatically proved its correctness. Regrettably this vindication came too late for Maxwell, who had died of cancer five years earlier, one week before reaching his forty-eighth birthday."

"I assumed that this is not the same Hertz who rents cars." Beth quipped.

"It could be a relative, for all I know. What Heinrich Hertz observed was that electric sparks or oscillations that he generated in one part of his laboratory induced identical electric oscillations in another electric circuit elsewhere in the lab. This other circuit, however, was quite some distance away and was not powered by any electric source at the time. Hertz had little difficulty proving that he had, in fact, observed the transmission of an alternating electromagnetic field from the oscillating electric sparks to the second circuit."

"Is this how electric discharges caused the frog's legs mounted on Galvani's garden fence to contract during lightning storms?" Beth asked excitedly.

"Very good deduction!" I beamed (because all teachers like to take credit for their pupil's brilliance). "Any electric oscillation, described by its *frequency* or number of cycles per second, emits an electromagnetic field oscillating with the very same frequency."

"Are radio waves also electromagnetic radiation?" Beth was encouraged to ask next.

"Yes they are. As it happened, what Hertz had observed was the transmission of a radio-frequency field to his passive receiver."

"I don't understand," Beth interjected, "isn't a radio receiver normally connected to an electric power source?"

"Yes it is, but this power is used to amplify the incoming radio signal not to detect it." I explained. "Although we are not aware of it, we are constantly being bombarded by electromagnetic radiations from all kinds of sources at all different frequencies. We call it the *electromagnetic spectrum.* This spectrum ranges from just a few cycles per second to the very high frequencies found in x rays and gamma rays."

"Is any of this radiation harmful?" Beth wondered. "I don't like the idea of being struck by anything I don't know about."

"For the most part, the evolution of the human species testifies to our ability to adapt to the presence of such radiation." I tried to reassure her. "As you know, excessive exposure to the ultraviolet rays in sunlight can cause sunburn and even skin cancer in extreme cases. Other radiations also may produce undesirable effects but don't think that they all come from people-made sources. The sun and more distant stars are constantly emitting electromagnetic radiations of differing frequencies that reach us on earth and from which we have no way of escaping."

"Tell me some good news."

"The good news is that the electromagnetic spectrum contains, in addition to the ultraviolet light, neighboring frequencies that provide light by which we can see each other, x rays that help our dentist identify teeth that need to be drilled, radio and television frequencies that entertain us in our free moments, and radar or microwaves to heat up our TV dinners so as to enable these moments of entertainment."

"Speaking of x rays, what exactly are they?"

"As you know, the discovery of x rays took place by accident." I responded. "Just about one-hundred years ago, physicists were studying the properties of electrons accelerated by electric fields inside evacuated glass tubes. Emitted by a negative electrode called a *cathode*, the electrons accelerate toward a positive electrode called an *anode*. Among the many studying this phenomenon was a relatively obscure physics professor at the University of Würtzburg in Bavaria, called Wilhelm Conrad Röntgen. Working in his darkened laboratory on November 8, 1895, he noticed that a fluorescent screen placed a couple of yards away from his tube would light up whenever he turned on its voltage. He quickly traced the origin of the mysterious radiation causing this fluorescence to a spot on his glass tube that was struck by the accelerated electrons inside it."

"Sounds like Röntgen was an alert detective!" Beth chimed in.

"He was indeed. Within two months, Röntgen had established that this radiation, which he named *x rays*, travelled in straight lines and cast sharp shadows, passed more readily through flesh than through bones, was even more readily absorbed by metals, and that it probably was a part of the electromagnetic spectrum newly described by Maxwell."

"All that in just two months when he had not a clue as to what these x rays really were!" Beth was truly impressed.

"Röntgen described all these findings in New Year's greetings he mailed to some of his colleagues. Shortly thereafter, he reported his findings to the Würtzburg Phisico-Medical Society."

"I bet that caused quite a sensation!" Beth was still impressed.

"The news spread like wildfire. Within one year, dozens of hospitals and laboratories had constructed x-ray tubes and, literally, hundreds of technical papers describing the generation, application, and characterization of x rays had been published around the world."

"One last question," Beth asked, "if other physicists had been working with similar electron tubes, how come they did not also detect the x rays?"

"They probably did." I explained. "We know, for example, that an English physicist, for whom they had been named *Crookes tubes*, had noticed that some photographic plates near his tube became inexplicably fogged. So he moved the plates to another cabinet and solved his problem."

"What a way to miss getting a Nobel Prize!"

"That's right." I added. "Röntgen received the very first Nobel Prize for physics in 1901."

# Chapter Ten:
# Breakfast of Eggs and
# Crisp Bacon

## *Making Waves*

"What exactly did Maxwell mean when he told his intended that he knew what light was?" Beth began at our next breakfast. "Before you answer that question, would you like some bacon with your eggs?"

"Yes, please. It looks nice and crisp; just the way I like it." I replied. "As to Maxwell, he knew that light is composed of an electric and a magnetic field that oscillate in unison and with the same frequency, as they move together through space at the fastest speed we know."

"I guess I am having trouble picturing such oscillating fields moving through space." Beth sounded puzzled.

"All right, try to visualize a tiny electric charge oscillating back and forth about some fixed point." I suggested, "Such an oscillating charge generates an electric field that starts from zero, when the charge is at its midpoint, and gradually increases to some maximum value, when the charge reaches its maximum displacement. When the charge next heads back in the opposite direction, the magnitude of this field decreases to zero and then proceeds to increases in the opposite direction (Fig. 19). After reaching its maximum value in the opposite direction, the field declines back to zero and repeats the cycle over and over again.

If you can picture one such field moving through space, attach to it a second magnetic field that oscillates back and forth at right angles to the electric field but otherwise looks and acts the same. According to Maxwell, both fields are produced simultaneously by the single oscillating electric charge and, together, make up the moving electromagnetic field."

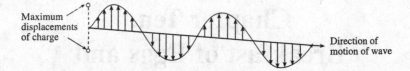

Fig. 19. The charge oscillating about a central point generates a succession of electric-field displacements represented by the arrows moving to the right. The magnitudes of these arrows increase from zero and track the charge's displacement back and forth as the total field produced (outlined by the solid curve) moves to the right.

"It seems to me that you are describing a wave-like motion in this electromagnetic field." Beth observed.

"That's quite astute." I noted happily. "Because the wave nature of light is so important to the rest of my stories, let me take a moment or two to describe wave motion in a bit more detail."

"Before you go on, is the bacon crisp enough for you?"

"Just the way I like it, crunchy!"

"Now, picture a stone being dropped into a still pond of water." I went on. "It momentarily displaces the water below the stone. The water reacts by welling up from below to restore the surface level once more. Such behavior is known as the *Le Chatelier Principle*, by the way, which states that there is a tendency in nature to restore the status quo or equilibrium that preexisted any disturbance."

"Now here, at last, is a principle I can approve!" Beth exclaimed.

"In actuality, when the displaced water surges back," I continued, "it disturbs the status quo that existed at the moment of its maximum displacement so that the resurgent water encounters a reaction that again depresses the water surface, not unlike a metal spring bobbing back and forth. The effect of this up-and-down motion of the water is that a circular wave radiates outward (Fig. 20) from the place where the stone fell in."

"I've made water waves like that by throwing pebbles into a pond more than once," Beth reminisced.

"There is an important aspect of this water wave that I want to point out. The work done by the stone in displacing the water initially is transformed into two kinds of energy. One is the energy needed to continue the up-and-down motion of the water column. The other is the

Fig. 20.   A stone dropped into a still pond produces circular waves spreading out in all directions.

energy that flows outward with the circular wave (Fig. 20). Note that, as the radius of the water wave increases, the energy it carries is spread over a growing circle and becomes less per unit distance along its outer edge. This means that less energy is available to raise and lower the water along the spreading wave so that the height or amplitude of the wave gradually diminishes or dies out."

"Does the wave carry water with it as it moves out?" Beth asked.

"No it doesn't." I was quick to point out. "The outward spreading wave acts to raise or lower the surface water that is already there in the pond. As the wave moves outward, it does so by raising the water at right angles to the wave's motion, without displacing water along its path. I am sure that you have seen a leaf bobbing up and down, as a water wave passes beneath it, without displacing the leaf at all. All this illustrates a very important point, namely, that the nature of a wave does not depend on the medium through which it moves or propagates, but the wave does need a medium to enable its propagation."

"Do electromagnetic waves need some medium for their propagation?" Beth asked. "If so, what is the medium?"

"Excellent question!" I responded. "Nineteenth-century physicists were convinced that light waves, like water waves or sound waves, had to have some medium through which to travel. So they created one of the strangest concepts in physics, that of an invisible and odorless *ether* that had no mass or density, so that it could not be detected by any means. Yet this ether was presumed to permeate all of space, even vacuum!"

"Doesn't sound any stranger to me than some of the other things physicists believe." Beth murmured.

"Actually the notion of an all-pervasive ether was philosophically appealing because it removed the strangeness of a vacuum devoid of anything imaginable. It also supported the Aristotelian notion that the cosmos contained no empty spaces anywhere. But appealing or not, the sole purpose of this ether was to transmit light waves!"

"You speak of the ether in the past tense. Is it no longer with us?"

"Its existence was disproved toward the end of the nineteenth century by a couple of clever experimentalists, Michelson and Morley, who actually set up their experiment initially to prove that an ether does exist." I explained. "Not surprisingly, physicists, including Michelson and Morley, did not accept their findings until the experiments had been repeated a number of times. Finally, Einstein administered the *coup de gras* in 1905 by showing that such an ether was not at all necessary for the propagation of electromagnetic waves, including light waves."

"What about sound waves," Beth wanted to know, "do they need some medium to travel through?"

"Yes, indeed," I replied." Only electromagnetic waves can travel through a total vacuum whereas sound waves need air or some other medium, liquid or solid, to propagate. In doing this, the sound waves actually displace the medium analogously to the displacement of a water surface by a water wave."

"Can you illustrate that for me? For instance, when I strike a key on the piano, what produces the sound I hear?" Beth asked next.

"As you know, the piano key is linked to a hammer. When you hit a key, its hammer strikes one of the strings on the piano's sound board. That causes the string to vibrate. The vibrations displace the surrounding air and set up a sound wave that spreads in all directions. In musical instruments," I went on, "this typically also sets a sound board into vibration, which can displace even more air because of its larger size. In this way, the sound is amplified and made louder."

"Is that why a grand piano sounds louder than a small upright?" Beth mused aloud. "And how do we make sounds when we talk?"

"We vibrate our vocal chords to produce sounds. What comes out is determined by the way we shape our mouths, breathe, strain our chords, and so on. Similarly, the sounds produced by a violin are affected by its shape, the wood from which it is made, how that wood has been finished, and many other variables that combine to produce the final product."

"When I play the piano, each key that I strike produces a definite musical note." Beth persisted. "To produce such notes in a string instrument like a violin or guitar, one has to place a finger on different parts of the strings. How do these two kinds of instrument relate?"

"If you look inside a piano, you will see strings of different lengths. Each of these produces one note, which is determined by the length of the string, the tension it's under, and its mass. Similarly, by fingering the strings in a violin, one fixes the length of the string that stroking by the bow sets into vibration and that controls the notes produced thereby."

"How do humans hear sounds?" Beth asked next.

"The frequency of the sound wave modulates the air through which it travels by making the air more or less dense along its path. These variations in air density affect the air pressure against our ear drums, in effect, making them vibrate with the same frequency as the sound wave. Our brain then recognizes these vibrations as sound."

"So when I hear you crunching away on your bacon, what I hear is your teeth making waves!"

## What Waves Can Do

"What's even more interesting is the way sound waves can interact with each other to produce echoes and various other sound effects that you often hear. Such interactions, by the way, are characteristic of all kinds of waves, not just sound waves."

"That sounds interesting. Please tell me how you characterize a wave." Beth requested.

"Good idea." I began. "A wave, in its simplest form, undulates in a regular fashion and repeats this undulation over and over again as it moves through its medium. You have probably tied a rope to a wall, at some time in your youth, and wiggled the loose end up and down to set up a wave pattern in the rope (Fig. 21)."

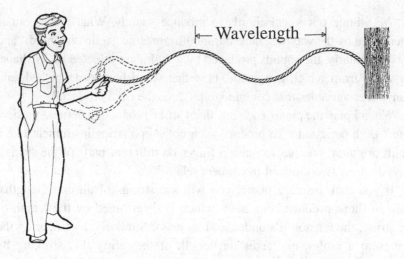

Fig. 21.   By moving the free end of the rope up and down with different frequencies, it is possible to set up waves having different wavelengths in the rope.

"I've done that!" Beth exclaimed. "The faster you wiggle the rope, the more waves are created in it."

"That's right." I continued. "The frequency with which you vibrate the end of the rope up and down, determines the number of complete waves appearing in the rope's displacement. We describe the waves produced by a wavelength, measured from one crest to the next. The wavelength is inversely related to the frequency of the wave. So, the more cycles, or complete wavelengths, produced per second, the shorter the wavelength of each wave."

"If I understand what you are saying," Beth interposed, "by shaking the free end of the rope faster, I increase the frequency and shorten the wavelength of the waves produced. It is possible, therefore, to describe a wave either by its wavelength or by its frequency, since one is the reciprocal of the other."

"That's about it, except to note that the height of the wave produced is determined by how far up and down you shake the free end. This is called the *amplitude* of the wave and represents the maximum displacement of the medium through which the wave travels at some specific speed."

"What happens when a wave reaches an obstruction like a wall?" Beth asked.

"That depends on the nature of the wave." I responded. "Most commonly, all or part of the wave will be reflected by the wall. If the wall does not block the entire path of the wave, then it may bend around the obstruction and continue its forward motion."

"We can illustrate this most easily by means of water waves." I went on. "When the leading edge of a wave, called its *wave front*, reaches an impenetrable wall, it will be reflected back so that its direction of travel is altered (Fig. 22(a)). When such a wave front reaches a small obstacle like a vertical post, it will flow around the obstacle while reforming the wave once it has passed the obstacle (Fig. 22(b)). The bending of a wave as it flows around an obstacle is called *diffraction*, a term we shall encounter repeatedly later."

"I am aware that the waves reflected by a wall meet the incoming waves so that their individual wavelets overlap." Beth observed. "Doesn't this cause some sort of interference? What does it do to the resulting mish-mash in the water?"

"Fortunately the intersecting waves create no mish-mash. Whatever that is." I couldn't help laughing at Beth's choice of language. "Keep in mind that a wave moves forward but the water does not. The surface of the water is merely displaced up and down as the wave passes through it."

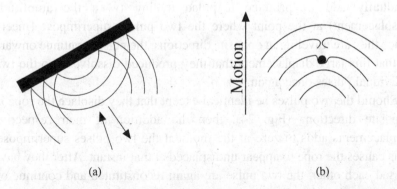

(a)                                    (b)

Fig. 22.   (a) A wave striking a wall is reflected so that it now flows in a symmetrical direction away from the wall. (b) As a wave passes a small obstruction, it reforms the wave front by flowing around the obstruction and continues its forward motion.

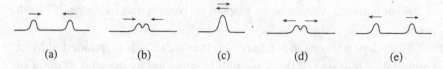

(a)              (b)              (c)              (d)              (e)

Fig. 23.   As two like pulses approach each other (a), their respective displacements add producing a combined pulse (b), that grows into a double pulse (c), before they separate (d), and go their separate ways (e).

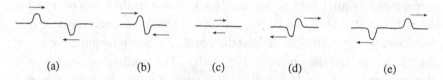

(a)              (b)              (c)              (d)              (e)

Fig. 24.   As two equal but oppositely directed pulses approach each other (a), their opposite displacements start to cancel each other (b), adding to zero (c), before reconstituting the original pulses (d), and they proceed to move off (e).

"It is easiest to illustrate what happens," I continued, "by considering a pair of like pulses (half waves) passing down a rope (Fig. 23). Each pulse can be thought of as a set of closely spaced displacements of the rope. As the two pulses approach each other, the displacements each pulse produces add together so that the amount of total displacement is, simply, their sum. If the two pulses are alike, the displacement they produce will gradually add to produce a pulse having twice the amplitude (displacement) at the point where the two pulses superimpose (meet). Since they are traveling in opposite directions, the pulses continue onward so that this large displacement that they produced dissolves into the two individual pulses once again.

Should the two pulses be identical except that they displace the rope in opposite directions (Fig. 24), then the addition of their respective displacements adds to zero at the moment the two pulses superimpose. This causes the rope to appear undisplaced at that instant. After they have passed each other, the two pulse are again reconstituted and continue on their respective ways."

"This also happens when two water waves meet?" Beth asked with some doubts lingering in her voice.

"Yes." I responded. "Called the *superposition principle* for waves, it holds equally for one-dimensional waves traveling down a rope, two-dimensional water waves, and three-dimensional sound waves."

"We call the reinforcement of the two waves (Fig. 23) *constructive interference* while the annihilation of the two waves (Fig. 24) is called *destructive interference*." I went on. "The same kind of interference can take place in two or in three dimension in exactly the same way. Thus what happens when two waves meet in a rope is no different than what happens to two water waves crossing paths in water or two sound waves in air."

"Do the same principles extend to the combined sounds that an orchestra or a choir can produce?" Beth wanted to know.

"The superposition principle of waves, that is, the ability to add the displacements each individual wave produces to each other, is what creates the glorious sounds as well as the discordant sounds we sometimes hear.

When listening to music in a concert hall," I elaborated, "you must further add the reflections of the sounds by the ceiling, floor, and walls of the hall. Since some of these have to travel a longer distance than the direct sounds from the sources producing them, there can be some destructive interference between the various waves reaching one's ear and this can distort the sound quality."

"Is that why I've seen baffles and reflectors scattered around the ceilings and walls of certain concert halls?"

"Right once more!" I couldn't keep the pleasure out of my voice. "That is also why sound-damping materials are used to line the inner surface of a room. These materials, like heavy drapes or acoustic tiles, absorb the sound reaching them or scatter the waves into a multitude of low-amplitude waves whose interference with desired sounds is minimized."

"One last question." Beth requested. "Is it possible to have two sounds interfere destructively so as to produce no sound at all?"

"It is certainly possible in principle. How easy it is to accomplish in practice depends on how well we can characterize the sound we would like to eliminate. With modern acoustic equipment and high-speed computers, this is becoming more feasible and some commercial devices have actually appeared on the market. Ironically, there are locations in some auditoria where this kind of destructive interference occurs unintentionally in what are called 'dead' spots in that hall. Anyone sitting in such a location has great difficulty hearing what's being said on the stage."

"Does any of this have a relation to the different way the siren on a police car sounds when it is moving towards you than when it is driving away?" Beth asked next.

"Maybe it's because you're more worried when a police car is coming after you than when it is going away?" I quipped. "Actually the cause of the different pitch you hear was first explained correctly a hundred and fifty years ago by an Austrian physicist named Doppler. To understand what happens, picture sound waves traveling radially outward from a source like the water waves we saw moving outward from the dropped stone in a pond (Fig. 20). At different moments, the circular wave front has grown to a different size. Now, suppose that the source is moving in some direction at a constant speed. As the source moves forward, the wave fronts it produces pile up closer to each other ahead of the moving source and spread further apart behind it (Fig. 25). Since the wavelength of waves is the distance between crests in successive waves, a person

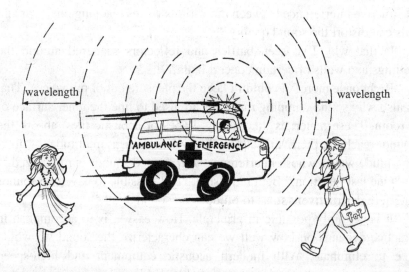

Fig. 25.   As the source of sound moves to the right, wave fronts emitted at successive points are more closely spaced ahead of the motion than behind it, causing a listener on the right to hear a higher pitched sound than that heard to the left.

standing ahead of the moving source hears sound having a smaller wavelength, hence higher frequency or pitch, than a person standing behind the moving source."

"What if the listener is also moving?" Beth wondered aloud.

"It is the relative speeds of source and listener that matter." I explained "If your car is moving past a stationary source, the pile up of successive wave fronts from a sound source will be exactly the same as when a stationary listener hears a moving source. In fact, the Doppler effect is what police officers use to monitor the speed of cars on the highway. Only they use inaudible radar waves instead of sound."

"So," Beth observed, "the radar trap that caught Harry speeding last month is another fall out from discoveries made by physicists a long time ago.

"Almost every bit of knowledge that has anything to do with the functioning of material things can be traced to some discovery in physics."

"Didn't you tell me last year that wild story about how your old fishing buddies used radar to detect fish?"

"Yes, I did," I remembered. "Although they admitted spending more time trying to get the system to work than in spotting fish with it. The fish scanner actually is an outgrowth of the underwater detection devices developed for submarine warfare in World War II."

"There is another outgrowth of war-time research and that is the jet airplanes that now fly at speeds exceeding the speed of sound." I went on. "That, in turn, produces an interesting effect, popularly known as the sonic boom."

"I've often wondered about that." Beth interjected. "I remember when military jets first started flying at such speeds, people living near military airports became very angry at the noise and the alleged damages that it created."

"There was an amusing incident following such complaints. The U.S. Air Force decided to disprove the claims that the noise caused picture windows and other fragile object to break. So they assembled a group of newspaper reporters at an underground bunker in Nevada, some distance from a cluster of make-believe houses intended to resemble a small city. After the reporters were outfitted with ear muffs and told to watch special meters in the bunker that were connected to pressure detectors inside these houses, the commanding general signaled a pair of jets to 'buzz' the model city. The reporters could see the planes roar past the town while the needles in the meters didn't even flicker. Encouraged by this, the general ordered the planes to makes several passes at ever lower altitudes. The meters still did not move. Convinced that jet aircraft do not damage buildings, the general loaded the reporters in a bus so that they could inspect the model town. What they found was that all the buildings had collapsed — leveled by the shock waves of the sonic booms!"

"Why didn't the meters register anything?"

"That's what a red-faced Air Force general is still wondering."

"What causes such a sonic boom?"

"You've seen a speed boat build up a bow wave as it knifes its way through the water." I responded. "A jet aircraft builds up a comparable three-dimensional wave in the air through which it flies (Fig. 26). In this case, the 'bow wave' is a pile up of compressed air along a three-dimensional cone trailing the plane, rather than a V-shaped bow wave of piled-up water that trails a boat. Once the jet's speed just equals the speed of sound, the crests of all the wavefronts superpose exactly, building up a single highly compressed cone of air. When such a cone of compressed air

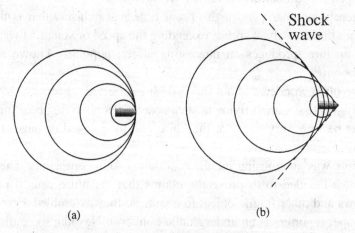

Shock wave

(a)          (b)

Fig. 26.   Build up of a shock wave (cone in three dimensions) by a fast moving object. (a) As its speed just equals that of sound. (b) After its speed exceeds that of sound.

passes a person on the ground, that person hears a sharp noise not unlike that produced by a shock wave of compressed air following a brief explosion of some kind. It is this shock wave that may break a window or do other structural damage."

"So, hearing a sonic boom on the ground is like being splashed by the bow wave of a passing speed boat while standing on a dock." Beth remarked. "Is this also related to breaking the sound barrier? Whatever *that is*."

"Quite so," I responded. "Once the jet reaches the speed of sound, the cone of compressed air presents a physical barrier to any further acceleration. Upon exerting additional energy, the plane can break through this compressed air or sound barrier (Fig. 26(b)). After that, the jet is said to be flying at a *supersonic* speed with the cone of compressed air trailing behind it."

"Is supersonic the same as ultrasonic?" Beth asked.

"No. Supersonic means larger than sound while ultrasonic means outside our ability to hear."

"So a supersonic plane flies faster than the sound it makes while an ultrasonic whistle can be heard by dogs but not by people." Beth asked, "is that right?"

"Yes, it is."

Beth's last words were: "Well, I must say that hearing all this *sounds* very interesting."

# Chapter Eleven:
# Breakfast of Oat Meal with
# Light Cream

## *What is Light?*

"This morning I am serving you hot cereal with light cream." Beth announced on a cold but sunny day.

"I'll bet you chose light cream because we agreed to discuss light this morning." I responded.

"You already told me that light is an electromagnetic wave. But how does that change what people thought light was before Maxwell?" Beth inquired.

"Maxwell made his discovery only a little over one hundred years ago. Other prople have been speculating about visible light far longer than that and physicists have been attempting to understand the various properties of light for three hundred years at least.

It's really a very interesting story." I continued. "Newton's friend and colleague, Robert Hooke, was advocating a wave model to explain some known properties of light, but Newton argued that light was actually corpuscular in nature. His reasoning was based on the fact that a candle illuminating a large wall will cast a sharp shadow if an opaque board shields part of the wall (Fig. 27). Were light a wave, Newton argued, it would be diffracted by the obstruction and bend around it the same way that a water wave does. In that case the shadow it casts should not be sharp."

Beth interrupted to ask: "What does corpuscular mean?"

"Corpuscles are particles, so that 'corpuscular' means 'made up of particles'." I responded before going on with my story.

106

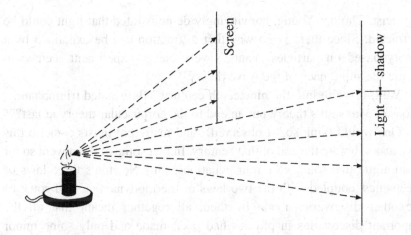

Fig. 27. The part of the wall illuminated by a candle is separated from the part shaded by an opaque screen by a sharp boundary indicating that the light travels to the wall in straight lines the way particles like bullets might.

"Newton developed a rather extensive theory in which he sought to explain the properties of light in terms of its corpuscular nature. He did this despite the much more elegant and simple explanations being put forth by a contemporary Dutch scientist using a wave model for light. Why don't two light beams intersecting each other collide destructively, Christiaan Huygens wondered, in the same way that two intersecting streams of arrows would? We know, after all, that waves can pass through each other but not particles."

"Here is another example of two expert scientists proposing totally conflicting explanations; and of such a very fundamental issue as the nature of light!" Beth marveled. "How was their disagreement resolved?"

"It wasn't for a long time. To Newton's credit, he recognized some of the difficulties inherent in his corpuscular theory and he did not, therefore, reject the notion of wave-like light entirely. Newton's followers were so awed by his fame, however, that they were much less tolerant in their views of competing theories so that the corpuscular theory of light dominated for the next one hundred years.

Gradually the corpuscular theory began losing supporters," I went on. "At the start of the eighteenth century, the English physician and physicist, Thomas Young, convincingly demonstrated that light could be diffracted. Since there is no way that diffraction can be explained by a theory based on particles, Young's very clever experiment seemed to assure the supremacy of the wave theory."

"Well, if we're into the nineteenth century," Beth stated triumphantly, "doesn't Maxwell's theory put an end to the corpuscular theory at last?"

"One would think so," I observed, "and many physicists came to this conclusion before the end of that century. In fact some of them went so far as to argue that Maxwell's four equations, plus Newton's three laws of mechanics, coupled with the two laws of thermodynamics, and some of the other discoveries made by then, all together meant that all the important discoveries in physics had been made and only some minor details remained to be filled in."

"Whenever one hears such smug pronouncements, it's a sure bet that something dramatic will come along to disrupt the complacency." Beth announced. "What did it this time?"

"Actually, it was not a single event," I reported. "It was more in the nature of several small shocks that began to suggest that, maybe, light was not really a wave after all. Most noteworthy was the observation made in 1899 by Philip Lenard that light falling on a metal plate ejected electrons from the plate, a phenomenon he named the *photoelectric effect*. Since electrons are particles, there is no way that energy carried by a light wave can be concentrated so as to eject an electron from the metal. A corpuscular light could, of course, in the same way that a marble directed at a pile of marbles can dislodge another marble from the pile."

"A fine dile—e—emma we—e—e have here." Beth sang from Gilbert and Sullivan's *Trial by Jury*. "How did physicists deal with this turn of events?"

"Not easily. Legend has it that on Mondays, Wednesdays, and Fridays, physicists believed that light was a wave, while on Tuesdays, Thusdays, and Saturdays, they believed it to be corpuscular."

"What did they do on Sundays?"

"Most of them probably went to church and sought divine guidance."

"I can't wait," Beth announced. "How was the dilemma resolved, or is it?"

"In 1905, Albert Einstein published three landmark papers. One of these argued that light was both — a particle *and* a wave, simultaneously!

What's more, Einstein used the photoelectric effect to prove this. It is ironic that 1905 is the year in which Lenard received the Nobel Prize in physics for having discovered the photoelectric effect six years earlier. Einstein, had to wait sixteen years to receive his Nobel Prize for explaining what made this photoelectric effect possible!"

"I know I shouldn't ask now, but what were his other two landmark papers?"

"One dealt with a phenomenon called the Brownian motion of submicroscopic particles and served to confirm the atomic theory of matter. The other was Einstein's theory of relativity, which completely revolutionized the physics of the twentieth century."

"I bet that didn't happen overnight," was Beth's ending comment.

## How and What We See

"A baby discovers very quickly that it needs light in order to see." I resumed our discussion of light. "The rest of us just take its existence for granted. The other day, a popular daily columnist in our daily paper claimed that we only see a light beam passing through the air because it illuminates the dust particles suspended in the air. Has he never looked at the source of the light directly? What makes the dust particles visible is their ability to scatter the light incident on them toward our eyes."

"How do the dust particles scatter light?" Beth asked. "Why doesn't the light flow right past them in the same way that waves pass around posts or other obstructions that they may encounter?"

"A full understanding of just how matter interacts with light only became possible after the major breakthroughs in physics that occurred in the twentieth century. We call this field *physical optics* to distinguish it from the much earlier, empirical observations by Huygens and others that make up the field of *geometrical optics*." I began to explain. "We now know that the electromagnetic wave called light can impart all or part of its energy to the atoms in matter and that these atoms then become sources of identical light waves going out in all directions from the atoms. But we can study the interaction of light with matter without necessarily understanding such interactions with atoms, and this is what the physicists did as they developed our understanding of geometrical optics."

"Can you give me some examples?"

"Gladly," I responded. "As far back as the first century A.D., Hero of Alexandria concluded that light takes the shortest path as it travels from its source to the eye of the observer. Actually this was corrected by the French mathematician Pierre de Fermat fifteen-hundred years later to state that light always follows the path that takes the shortest time. This is not an important distinction for us because, either way, it leads to the conclusion that light striking an object at some angle is reflected back to us at exactly the same angle (Fig. 28).

Other geometrical relations regarding the path followed by a light beam came later. For example, light passing from one medium to another is bent from its original path. That is why, when you look at your foot while standing in a lake, it appears to be closer than it would if you look down at it on the shore. This bending of light is called *refraction*.

"Is there any practical application for geometrical optics?" Beth asked with renewed interest in her voice.

"Quite a lot." I responded. "Even before Hero's theorem, the reflection of light by bright surfaces was well known. Thus Archimedes advised the construction and placement of concave mirrors that focused the sun's rays on the sails of incoming enemy ships. By this means he was able to set the sails on fire and save his home town of Syracuse from an invading fleet.

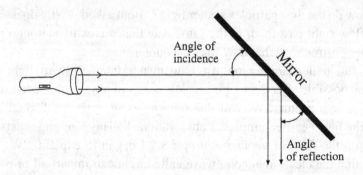

Fig. 28.   A light ray from the source at left reaches a mirror at some angle and is reflected to the observer at the same angle because only then will the two straight-line paths sum to the shortest possible total path.

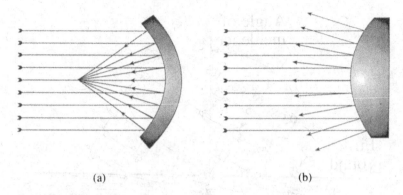

(a)                              (b)

Fig. 29.   (a) A *concave* mirror reflects parallel light rays toward a common center. (b) A *convex* mirror reflects incoming parallel light rays away (disperses).

A concave mirror, by the way, is one whose curvature is such that parallel rays approaching it are bent toward the center of curvature (Fig. 29(a)). A convex mirror, by comparison, bends such rays away from the reflecting surface (Fig. 29(b)). All this, in full accord with Hero's or Fermat's principle of reflection."

"Refraction," I went on, "is responsible for the operation of lenses, like those in eyeglasses, magnifying glasses, and space telescopes. Because glass is denser than air, light rays entering the glass are slowed down and this is what bends them from their path in air. As you can imagine, convex-shaped glass will bend the light rays one way, while concave-shaped glasses will bend it another way. The result is that the light rays are concentrated in different regions after passing through the glass."

"Hold up a minute." Beth pleaded. "Why should the light bend from its path because it is slowed down?"

"Consider a troop of boy scouts marching across dry land. Suddenly they encounter a very muddy swamp which slows down their progress. Suppose the parallel rows of scouts approached this swamp at some angle (Fig. 30). The first scouts to enter the swamp are forced to slow down. Since those on the dry land keep up their original pace, the intervals between rows shorten as they enter the swamp. This change in the spacing between rows has the effect of bending the parallel rows of scouts at the point where they cross the boundary between the dry land and the swamp.

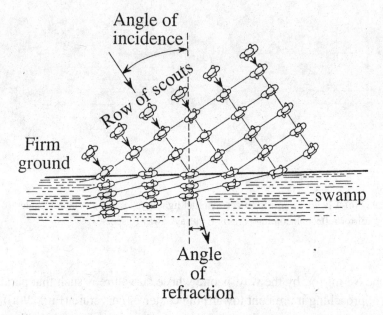

Fig. 30.  Refraction, or the change in speed, of parallel wave fronts (or rows of marchers), causes the direction of propagation to bend toward the perpendicular to the interface in the slower medium as compared to the faster (or less dense) medium.

As the spacing between the rows decreases, their common direction of forward motion changes from the original one."

"Instead of rows of scouts, think of parallel wave fronts of visible light undergoing the same kind of change. This is a sample of the kind of analysis Huygens proposed for the refraction of light waves. It was so successful that it is still used in teaching geometrical optics to students on this very day."

"By bending the light rays, either by reflection or refraction," Beth observed, "it is possible to change the way something looks, isn't it? That's why, for example, they place curved mirrors in fun houses that can make one look skinny or fat."

"You're right on target!" I exclaimed happily. "The matter of forming images is what geometrical optics is all about. It is not a difficult subject, but it requires a lot of careful drawings to illustrate the many different possibilities. It does not lend itself too well, therefore, to a discussion at the breakfast table."

"I wish I could lose as much weight as my appearance in a curved mirror might suggest!" Beth wistfully announced.

"Is that what they mean when they talk about having done something with mirrors?" I wondered. "What we see in curved mirrors is called a *virtual* image and not the actual image of the object being viewed. The same is true of virtual images seen through lenses, which is why we can use lenses to magnify objects we no longer can see with naked eyes."

"Just how does an eye see an object?"

"An eye is a remarkable instrument whose basic design resembles that of a fairly simple camera but whose physiology is considerably more complex."

"I know the basics of how a camera works." Beth interposed. "Behind the lens in the front of the camera's case there is an opaque screen containing an adjustable opening. This diaphragm is combined with a shutter that can be opened or closed manually to admit light, as the photographer wishes. The light admitted then forms an image of the object being photographed on a film placed against the back of the camera case. Both the diaphragm opening and the length of exposure can be controlled by the photographer manually or by light meters built into newer cameras. Similarly, the distance from the lens to the film can be varied to focus the image on the film's surface.

I know that the eye also has a lens and a diaphragm called the iris," Beth continued, "and our eyelids act like shutters. The retina at the back of the eye serves as a film to record the image formed by the eye's lens. But I do not know how an eye can focus an image nor exactly how the retina communicates what it sees to the brain."

"The outer eye actually consists of a convex *cornea* whose function is to bend the incoming light rays toward the center of the eye." I took over. "Behind it, and embedded in an aqueous fluid, lies the *iris*, whose color can range from dark brown to light blue or bluish green, depending on the kind of pigments it contains. The iris does, indeed, control the amount of light it admits through the dark spot at its center, called the *pupil*, so that it does resemble the function of a camera's diaphragm. The pupil dilates in the dark and constricts in bright light. Unlike the shutter in a camera, however, the pupil also responds to the stimuli reaching our other senses. Thus the pupil tends to enlarge whenever we receive pleasant stimuli."

"Yes, yes," Beth couldn't restrain herself, "supposedly that's how one can tell when even an experienced card player holds a winning combination of cards. The pupils expand to display the pleasure. In the same way, contracting pupils can provide a warning that the person is feeling threatened or hostile."

"Directly behind the pupil is a transparent lens which focuses the image seen onto the *retina*." I resumed my description of the eye (Fig. 31). "It is held in place by delicate threads that are attached to muscles that can flatten or bend the elastic lens to adjust it visually to objects that are, respectively, near or further away. As a person ages, the lens becomes progressively less transparent and more rigid. This makes it more difficult to adjust the lens so that the person's ability to see clearly declines."

"But how does the retina interpret the light that falls on it to signal an image of the object seen by the eye?" Beth was impatient to know.

"The retina consists of many parallel layers of highly specialized and interconnected nerve cells. One of these layers contains rod-shaped and cone-shaped cells which act like miniature antennas in responding to different frequencies of the incoming light. Because the human eye evolved in sunlight, it is especially responsive to the frequencies in what we call the visible-light range of frequencies. The retina also can distinguish a wide range of relative brightness before transmitting all the signals it receives to the brain through the *optic nerve*. The spot where this nerve joins the retina contains neither cones nor rods, by the way, so that there is an actual blind spot in the retina that is insensitive to light rays striking it."

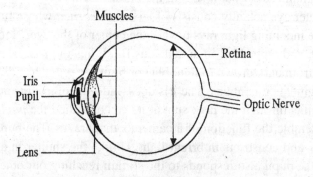

Fig. 31.  The human eye.

"At the risk of telling you more than you really want to know," I added as an afterthought, "the retina responds differently to the different colors or frequencies of light striking it."

"What do you mean when you talk of different frequencies making up the visible-light range?" Beth wanted to know.

"To answer this question we have to go back to Newton, who passed a narrow beam of sunlight through a prism made of glass (Fig. 32(a)). What he observed was that the white sunlight broke up into a *spectrum* of colored lights ranging from red to violet. He correctly concluded that the different colors making up white light traveled through the glass at slightly different speeds so that they were refracted by different amounts and emerged from the opposite side of the prism at slightly different angles."

"If sunlight, or what you call white light, is made up of different color lights that are refracted differently by glass," Beth now asked, "why do we see white light passing through glass windows?"

"Newton anticipated your question by placing an identical second prism next to the first (Fig. 32(b)), except that he turned it upside down. The dispersed color spectrum now strikes the second prism at the appropriate angles to be refracted by it back to reconstitute a white beam of light emerging from the opposite side," I explained. "This shows, incidentally, that white light is truly composed of the colored spectrum because we can recombine all these colors to recreate white light."

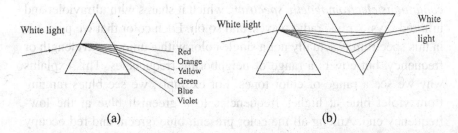

Fig. 32.   (a) White light entering a glass prism at some angle breaks up into spectrum of colors as it emerges from the opposite side. (b) If an identical prism is inverted next to the first prism, the colored lights enter it at just the right angles to be refracted back so as to reconstitute the white light emerging from the opposite side.

"I see!" Beth exclaimed. "If we push the two inverted prisms together, they will form a glass whose opposite sides are parallel. Does that mean that white light breaks up into colors inside the glass but re-emerges as white light on the outside?"

"The answer is yes." I replied. "To actually see the colors making up white light we need a transparent medium that has nonparallel opposite sides."

"Like the facets in our cut-glass vase or in the diamond ring you gave me so many years ago." Beth observed. "By the way, what is ultraviolet and infrared light that we hear about from time to time?"

"You recall that the term 'ultrasound' referred to sound whose frequencies were too high for the human ear to detect," I responded, "well, 'ultraviolet' is the continuation of the color violet whose frequency is too high for the human eye to detect. Similarly, 'infrared' refers to light below the red end of the visible spectrum that the eye no longer can see directly. When we bask in the sunshine, infrared light is absorbed by the atoms in our tissues and serves to warm those tissues, while ultraviolet light is absorbed by the cells in our skin and makes them darker."

"Are you going to tell me now how the bountiful sunlight makes it possible for us to enjoy all the beautiful colors that surround us?"

## It's a Colorful World!

"Thanks to Maxwell, we now know that white light is part of the *continuous electromagnetic spectrum*, which it shares with ultraviolet and infrared rays, x rays, radio waves, and so on. Each color that we perceive in this spectrum is actually not a single color with a unique wavelength or frequency, but covers a range of neighboring frequencies. This explains why we see a range of color tones. For example, we see blues ranging from violet blue at higher frequencies to a greenish blue at the low-frequency end. Among all the color present, blue, green, and red occupy the widest range in the spectrum so that they are known as the dominant or *primary* colors."

"What causes an object to have its particular color?" Beth asked. "And don't tell me it's the color of paint put on it."

"I wouldn't dare!" I chuckled. "When electromagnetic radiation falls upon any matter, one of three things may happen: all the incident radiation may be absorbed by that matter, in what we call an ideal black body. Or, conversely, all the incident radiation is re-emitted by the atoms in the irradiated body. To an observer, this appears like a scattering of the incident radiation uniformly in all directions. Most commonly, part of the radiation is absorbed and the rest is scattered. Because the energy of the radiation is directly proportional to the frequency, different colors of white light interact differently causing white bodies to scatter light of all frequencies, green bodies to scatter predominantly the green frequencies, and so forth."

"So green paint contains pigments that scatter green light." Beth repeated. "But what happens to the rest of the colors when white sunlight falls on the painted surface?"

"Colors other than green are largely absorbed by green paint." I explained. "That is why green leaf will appear to be black if viewed in red light, for example. Let me illustrate this for you by means of this small American flag. When viewed in white light, its red, white, and blue colors are clearly visible (Fig. 33(b)). Now look at it in red light (Fig. 33(a))."

"Oh! The red stripes still look red but so do the white stripes and stars, in fact, I can't tell them apart. But why does the blue field look black?"

"Because the red light illuminating the flag contains no blue light that the blue can reflect." I explained. "If we now view the flag in blue light (Fig. 33(c)), it is the red stripes that will appear black while the blue and white areas can reflect the blue light so that they appear to be blue."

"Oh, I see!" Beth exclaimed. "That is why colored clothes may look different in artificial lights than they do in sunlight."

"Quite so. Neon lights, for example, tend to predominate in the violet-blue end of the spectrum and this can make red colors take on a purplish hue." I continued. "Another kind of example is provided by the color of the sky. Because the nitrogen and oxygen atoms making up air scatter blue light much more efficiently than any other colors, the scattered light reaching us from the air molecules is predominantly blue. Any water vapor present in the air tends to absorbed infrared radiation which is partly the reason why rainy days tend to be cooler."

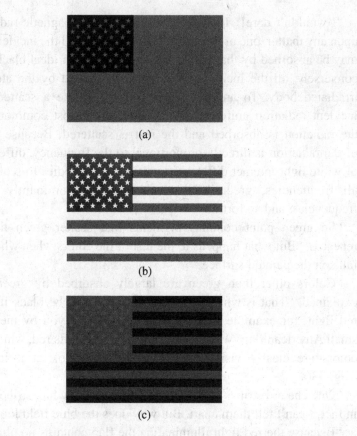

Fig. 33. Appearance of a United States flag when it is illuminated by: (a) Red light. (b) White light. (c) Blue light. (Photographed by John Hall, Institute of Materials Science, University of Connecticut.)

"But why is the sky so much bluer in the Caribbean Sea or on a Greek island in the Mediterranean than it is even at the sea shore in America?"

"Any particles or other molecules present in the air will cause different colors to be scattered or absorbed so that the color of the air or sky will be changed. As you know, our automobiles and industrial plants spew a variety of molecules and dust particles into the air so that we rarely get to see the true color of the sky any more."

"Is that why we're so concerned with the amount of ozone left in the upper atmosphere?" Beth excitedly asked. "The ozone molecules must be absorbers of harmful ultraviolet rays that radiate from the sun."

"That is precisely why it is essential to preserve the layer of ozone molecules intact. Unfortunately, not all of the causes destroying parts of this layer originate with humans. Nevertheless, it is another part of our environment that indiscriminate use of synthetic products helps destroy."

"If the sky is blue because air molecules scatter the blue part of the visible spectrum most efficiently, what happens to the red light at the other end of the spectrum?"

"The red light is scattered too but much less of it is scattered so we aren't as aware of it. This is further abetted by the fact that the cells in the human retina are more sensitive to the blue light." I explained "One effect of the preferential scattering of blue light appears as the sun sets." I continued. "When the sun gets close to the horizon, its light rays have to travels a longer distance through the atmosphere to reach an observer than while the sun is overhead. During this extra long path, most of the blue light is scattered so that the remaining light is mostly red. Which is why the sky appears to turn red at sunsets."

"And the sun itself, being the brightest object in the sky, also becomes the reddest at sunsets!" Beth announced proudly.

"Well, if you're so smart," I said teasingly, "can you explain why rainbows sometimes appear on a rainy day?"

"Hmm," Beth paused to think. "Rain consists of water droplets which must refract different colors differently, not unlike a glass lens. That would explain why they disperse colors like a prism. But it doesn't explain why we see rainbows rarely and not every time it rains. Why is that?"

"What happens is that a raindrop tends to disperse the component colors of white light at angles inclined by 40 to 42 degrees from the incident light rays. To see a rainbow, therefore, it is necessary for the position of the sun and the observer to be such that sun light refracted at these angles by the many rain drops in the air will actually reach the observer."

"Although rainbows must have existed forever, when was their origin first explained?"

"As a matter of fact, the first correct explanation was put forth by Isaac Newton around 1630."

"How could Newton explain a rainbow using a corpuscular theory of light?"

"Not easily." I responded. "Newton was forced to use some pretty convoluted arguments to explain any refraction of light based on his corpuscular hypothesis. Nevertheless, he did make many important contributions to our understanding of color phenomena, by being the first to observe them under controlled conditions in the laboratory and to analyze them carefully. Because of the profound effect his unifying theories of mechanical motion had on scientific thinking for many, many years, people sometimes forget that Newton was a first-rate experimentalist as well. For example, he was the first to demonstrate that a white object appeared to be white because it reflected all the colors of the rainbow equally. Similarly, he showed that a black object appears black because it absorbs all colors of the rainbow equally.

It was, however, that other giant of physics, Maxwell, who showed 150 years later that one could synthesize white light just by combining the three primary colors: blue, green, and red. Today we make use of that discovery in our color television sets which recreate all possible colors by a judicious mixing of these three colored lights."

"I remember from my early attempts to become a painter," Beth observed, "that it is possible to mix two colors to produce a third. Are you suggesting that I could have gotten by with just three colors on my palette?"

"Different criteria apply to blending colored lights than do to mixing colored paints," I explained, "because we see paints by preferential reflection of some colors and absorption of others by the paint. Moreover, the recognition of a color is a very subjective matter and can vary from person to person."

"That's for sure! You and I have argued more than once about a particular color being this or that."

"Have you ever noticed that a white table cloth looks as white in sunlight as it does in yellowish candle light?" I asked, determined to have the last word. "Can it be that our brains might be prejudiced?"

# Chapter Twelve:
# Breakfast of Lox and Bagels

## *What's the Speed of Light?*

"This morning I'm serving you lox and bagels." Beth announced one Sunday morning. "When I was growing up, my relatives would come to our house for Sunday brunch featuring lox, smoked whitefish, bagels, and cream cheese. So, in their honor, you are getting a special treat."

"That's great," I reacted with pleasure. "Hmm. Relatives. Relatives. I know. This might be a perfect time to discuss relativity with you because it ties in with our last discussion about light. You see, relativity is concerned with how the speed of light appears to different observers."

"When we say that something is 'relative,' don't we also have to indicate 'relative to what'?"

"Quite so," I replied. "For example, physicists like to make use of a reference frame consisting of two reference axes in two dimenisons and three reference axes in three dimensions We usually choose these axes to be at right angles to each other in what is known as the *Cartesian coordinate system.* This was first suggested by Descartes and so we named it after *Cartesius*, which is the Latin name for Descartes."

"I wasn't aware of the history," Beth remarked, "although psychologists also make use of Cartesian reference frames frequently to construct graphs showing how a behavior may change with time, frequency, or other parameters. I bet most people are familiar with reference frames even though they may not know their proper name or origins."

"Reference frames are particularly useful when we want to describe a direction to someone, or the motion of a body through space." I observed. "What happens, however, when the reference frame we are using is also moving? Consider a train traveling on earth at a speed of 50 miles per hour.

121

The earth's surface, however, is rotating about its axis at a speed of about 1,000 miles per hour, while its center is moving around the sun at a speed of nearly 70,000 miles per hour. So what is the actual speed of the train?"

"For people on earth," Beth was quick to respond, "it is, obviously, 50 miles per hour."

"That's right, but physicists tried for a long time to identify a reference frame to which they could refer any event within the known universe." I pointed out. "But the identity of such a reference frame continued to elude them. The reason for that became clear only after Einstein developed his theory of relativity."

"That certainly made quite a splash. Didn't it? And I seem to remember that Einstein wasn't even a very good student in school!" Beth observed. "I bet he was being bored by what his teachers were teaching in the same way that bright kids may be today."

"His poor grades did not discourage Einstein from applying to the most prestigious technological institute in Switzerland." I noted. "Much to the surprise of his former teachers, he was admitted after a second try."

"Perseverance wins again!" Beth exclaimed. "I'd like to understand his theory of relativity or do you think it's too complicated mathematically for me to grasp?" She continued.

"Actually, the concepts underlying the special theory of relativity are quite simple and can be understood by anyone who had an algebra course in high school. And I'm talking about the mathematical results of this theory!

The reason everyone speaks with such awe about it is that the theory of relativity leads to predictions that seem to defy common sense. Yet we now have abundant proof that, no matter how bizarre the predictions of Einstein's theory may seem in terms of our everyday experience, they have been verified repeatedly under widely different circumstances."

"Didn't Einstein once say that our common sense is nothing more than the collection of prejudices that we have acquired by the time we are in our late teens?"

"I believe he did." I replied and went on: "The easiest way to deal with his theory is to accept the two basic postulates that Einstein made in 1905: The first is quite innocuous. It simply states that the laws of nature remain unchanged as long as all reference frames are moving uniformly, that is, as long as none of them are accelerating. The second, however, insists that the speed of light is constant in free space and has the same value for all observers regardless of their relative motion."

"Why should there be a problem with accepting a constant speed of light in free space, by which, I assume, you mean in vacuum?" Beth asked. "Physicists had no trouble accepting Galileo's postulate of a constant gravitational acceleration on earth."

"It's the second half of that postulate that causes the difficulty, namely, that the speed of light remains constant regardless of whether the observer is moving or not." I explained. "We know from our daily experience that two cars heading south on the same four-lane highway can ride side-by-side at a speed of fifty miles an hour. If they happen to be going in opposite directions at that speed, they will pass each other at twice fifty or one-hundred miles per hour. Now consider a traveler moving in the same direction as a beam of light. According to Einstein, no matter how fast the traveler is moving, the light beam continues passing her at the same constant speed! If that doesn't shake you up, reverse the traveler's direction and increase her speed to nearly that of light. According to Einstein, the light coming toward the traveler still passes her at the same constant speed!"

"I am beginning to see what you mean." Beth observed in a hushed voice. "The only way that can make sense is if Einstein somehow changed the way we must measure distance and time, which define what speed is."

"Actually, the more important consequence of Einstein's second postulate is that two observers moving past each other at a constant speed can't tell which one is moving! In other words, only their relative motion has any physical meaning, hence the name, *theory of relativity*. You realize, of course, that this also eliminates the need for referring their respective motions to some universal reference frame."

"Wow!" Beth exclaimed. "This is going to take some getting used to."

## Is it Really Relative?

"Consider an open boat moving past an observer on shore at some constant speed (Fig. 34). Next, let a person standing on the port side, or

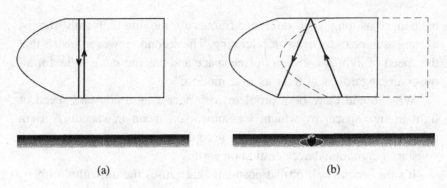

(a)                                        (b)

Fig. 34.  The way a round trip looks aboard a moving boat (a) is different from the way it is seen by an observer on shore (b).

side closest to the observer on shore, walk across the width of the boat to the starboard or opposite side of the boat and then back to the port side. Anyone aboard the moving boat will see the back and forth walk as a retracing of the same path (Fig. 34(a)) whereas the person on shore sees a longer path because the boat is moving past the stationary observer during the time it takes for the round trip (Fig. 34(b)). Now, the speed of the walker aboard the moving boat is determined by twice the length of the path from the port to the starboard sides divided by the time it took to complete the round trip. The person on shore, however, having recorded a longer path traced out during the same time interval (Fig. 34(b)) calculates a larger speed for the same walker. These two differently calculated speeds of the same walker can be easily reconciled by making an allowance for the speed of the boat. All this can be done using simple algebra."

"Are you hinting that even I could do it?"

"I am sure that you could do it," was my response. "But now, let me change the situation slightly. Suppose, instead of walking, the person aboard the boat sends a flash of light from the port side to a mirror on the starboard side which reflects it directly back to the port side. The observer on board the boat and the one on shore still observe two different pathlengths but, according to Einstein's second postulate, they must measure the same speed for the light! The inescapable conclusion, therefore, is that clocks that move keep time differently from clocks that stand still!"

"Wait a minute," Beth interrupted in an agitated voice. "How can clocks keep different times during the same time? That makes no sense."

"I did warn you at the outset that the conclusions from Einstein's theory will seem bizarre." I said soothingly, as I tried to calm my wife. "The reason they seem to be nonsensical is that we cannot observe the minuscule changes involved unless the two observers are moving past each other at a speed that, itself, approaches the speed of light. Thus we are totally unaware of the dilemmas presented by Einstein's theory in our daily lives."

"So how can we tell that it is valid?" Beth challenged.

"If you hang in here, I'll tell you how." I responded. "But, for the moment, keep in mind that we can reconcile the two clocks using nothing more than simple algebra. On the other hand, the theory of relativity also states that you can't tell who is moving, the observer on shore or the one on the boat — only that they are moving past each other. Therefore, whose clock needs adjusting?"

"I could say something, but I'll restrain myself." Beth stated with a big smile on her face. "By the way, would you like some more of this delicious lox?"

"It is indeed delicious, but I need to limit my intake of salty foods." I responded. "Coming back to the clocks, I do want to stress the fact that clocks moving very rapidly do keep time differently. This is not an illusion but a fact. It was verified using two jet planes carrying extremely sensitive and precise atomic clocks a few years ago. One jet was flying from East to West, opposite to the rotation direction of the earth's surface. The other was flying with the earth from West to East. As predicted, the times measured aboard the two planes were different. When they were compared to a similar clock at our Naval Research Laboratory near Washington, D.C., these tiny differences agreed exactly with the predictions of the theory of relativity."

"I still have one conceptual problem." Beth persisted. "If we imagine a vehicle moving past another at a speed approaching that of light, while we measure the speed of a light pulse aboard one of them, how can we be sure that the yardstick we used to measure the distance is not changing in its length instead of the time keeping by the clock?"

"As a matter of fact, both undergo algebraically identical changes." I responded. "This is why Einstein stressed that time and distance cannot be considered separately from each other. Instead, he showed how time and the three-space coordinates were all part of the same *space time continuum.*"

"Is this why time is called the fourth dimension?" Beth wondered aloud. "I wonder what my relatives would have made of this conversation had they overheard it back when I was a little girl."

## The Paradoxes of Relativity and Black Holes

"There is no question but that the theory of relativity poses some real paradoxes." I observed.

"I'm pretty sure I know what a paradox is, although right now I am no longer sure of what I am sure." Beth lamented. "Just how are you using this term?"

"A paradox is a proposition that appears to be true while seemingly contradicting usual beliefs or common sense." I resumed. "The ancient Greeks were very fond of paradoxical problems. In one of them, the fleet-footed Achilles gives a turtle a lead and then tries to catch and surpass the slower moving turtle. By the time Achilles reaches the point where the turtle started, the turtle has moved forward some distance. When Achilles reaches that point, the turtle, of course, has moved forward somewhat. The same holds true when Achilles reaches each point of the turtle's progress. According to this reasoning, which is without fault, Achilles can never catch up to the turtle!"

"I remember hearing that one before." Beth interposed. "Isn't that known as a paradox of Zeno the Elder? It sounds reasonable but it is based on a false premise that motion proceeds in a stop-and-go manner instead of being continuous."

"That's essentially true about all of Zeno's many paradoxes. They may be amusing brain teasers but they are all based on some incorrect premise." I agreed. "Einstein, recognizing the difficulties facing experimentalists trying to test his theory in the years following 1905, also liked to invent seemingly paradoxical questions to see if he could disprove any aspect of his theory. In one such *gedanken* or thought experiment, Einstein has one of two identical twins leave on a round-trip voyage into outer space while the other remains behind on earth. If the space traveler moves with a constant speed approaching close to that of light during a round trip, Einstein's theory predicts that she will age less than her twin sister who remained on earth. But, Einstein persisted, should both twins be kept blindfolded, they would both claim to have been the one that moved, since motion, after all, is relative."

"Quick, which one really turned out to be the younger?" Beth asked with a twinkle in her eye. "I want to do what she did."

"This paradox puzzled physicists for many years." I replied. "It can be resolved by taking into account the fact that the traveling twin has to undergo a very drastic acceleration to reach a speed close to that of light and an equally drastic deceleration before she can meet her older sister on earth. When proper allowance for these changes in speed is made, not only does the traveler turn out to be younger, but she is only too aware of the discomforts she must have felt at the changes in speed."

"This reminds me of some television footage showing test pilots undergoing huge accelerations. The horrible contortions seen in their faces make me wonder if even the pain of having a face lift may not be preferable."

"The best way to age is naturally and gracefully." I observed. "But I promised to describe to you the experimental verification of Einstein's theory. We now know that there exist subatomic particles called *mu mesons* which can be generated in particle accelerators on earth or by cosmic rays striking the earth's upper atmosphere. The mesons generated in accelerators usually travel with a speed of about one-tenth that of light and their average lifetime before they disintegrate is one-millionth of a second. Those generated in the outer layers of our atmosphere travel much faster but have an average lifetime ten times shorter. If we calculate the time it would take such a mu meson to reach the earth using their known speed and distance above the earth, it turns out to be nearly ten times longer than their lifetime. Thus none of these outer mesons should be able to reach us. Yet we observe such mu mesons reaching the earth all the time! We can resolve this apparent paradox by using the relativistic correction for the time predicted by the relativity theory."

"Are you saying that time advances more slowly for anything moving with nearly the speed of light?" Beth tried to understand. "Is that how the moving twin stayed younger?"

"Exactly right!" I replied. "And to confirm it, the mesons created in an accelerator on earth were divided into two groups moving at different speeds. Again, the faster moving mesons outlived the slower ones by the exact amount predicted by Einstein's theory."

"Is Einstein telling us to move as fast as we can to prolong our lives? Or is it just true that the faster we move the more ground we cover?" Beth mused aloud. "And what has become of Newton and his theory of gravity?"

"The gravitational theory is still with us and working well." I responded. "In fact, twentieth-century physics adopted a *correspondence principle* that requires any new theory to incorporate within it any correct theory that it replaces."

"What about space ship travel?" Beth asked next. "Does it have to be concerned about relativistic changes?"

"No and yes, is my paradoxical answer." I replied playfully. "Space ships do not travel fast enough to be affected by relativity concerns but they do travel in gravity-free environments and that may illustrate a further development made by Einstein."

"Consider an astronaut I shall call Jean," I continued, "moving at a constant speed far away from the intense gravitational fields of any planets. As we have witnessed in television footage of our astronauts, Jean is weightless out there so a ball that she releases from her hand stays afloat where she releases it. Should the rockets at the base of the spaceship be ignited at that moment, however, Jean and the ball will feel the acceleration to the right so that they will 'fall' toward the ship's floor on the left. In fact, if this acceleration matches that due to gravity on earth, Jean will feel and her ball will respond just as if they were feeling the force of gravity pulling on them, as on earth."

"Is this acceleration going to make Jean any younger?" Beth was quick to ask.

"At that acceleration, Jean will not be able to distinguish between life in her spaceship and what it was when the ship was sitting on its launching pad on earth." I resumed. "Suppose next that a meteorite enters the

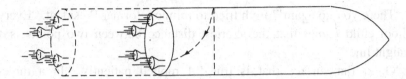

Fig. 35.  A meteorite penetrating one side of a spaceship accelerating toward the right follows the indicated curved path within the ship. To an observer inside the spaceship, therefore, it appears to be 'falling' toward the ship's floor.

spaceship through a side wall while it is accelerating. This meteorite will follow a curved path toward the floor (Fig. 35) just like a ballistic missile might on earth. On the other hand, an observer watching this event from a distant star, would see the meteorite continuing on its straight-line path while the spaceship's floor was rushing to the right to meet it. Which is the correct description of the meteorite's path? The curved one seen by Jean or the straight line seen by the alien observer on a star?"

"Assuming that Jean and the spaceship survived, I'll vote for our good Jean over some alien any time!" Beth declared loyally.

"That's a wise choice," I concurred, "because, once you take the spaceship's acceleration into account, the mathematics of the situation becomes inordinately more complicated. In fact, when Einstein set out to expand his original or 'special' theory into a 'general' theory of relativity, he had to invent the necessary mathematics because it simply did not exist before then."

"Didn't Newton also have to invent calculus in order to derive his laws of mechanics?" Beth interrupted to ask.

"Yes he did. Although Newton shrewdly did not use anything more difficult than algebra and plane geometry to prove his theorems when he presented them in the *Principia*."

"Whereas Einstein had only considered relative motion at constant speeds in his original treatment," I continued, "he now realized that it would be necessary to include such effects as gravity in his space-time continuum. This produces a number of unusual consequences, including the result that the shortest distance between two points may no longer be a straight line."

"There you go again!" Beth tried to mimic a former President. "Every school child *knows* that the shortest distance between two points is a straight line!"

"On a flat surface that is true." I replied patiently. "In a three-dimensional world, however, that simple rule has to be modified. Consider a plane flying from New York to Tokyo. Because the surface of the earth is spherical, if the plane tried to fly in a straight line, it would collide with the curved earth. So, instead, it follows a curved path called a *geodesic*. On earth, we call this a great circle and it happens to be the shortest path connecting any two points on the earth's surface."

"Is there any experimental proof to support Einstein's general theory of relativity?"

"Not only is there direct evidence that massive stars like the sun can bend the straight-line path of light rays passing near it, but we now know that even more massive bodies exist in outer space whose enormous gravitational attraction deviates any light rays passing by them. Actually, Einstein's theory found immediate application in explaining tiny deviations of the planet Mercury from the path predicted by Newtonian mechanics. More importantly, in time, scientists began to realize that our concepts of the universe would have to be modified. This has led to numerous new discoveries about how our universe began and where it is headed."

"If we are into discussing cosmology," Beth interjected, "what did Einstein have to say about black holes, whatever they are?"

"I am not aware that Einstein considered black holes directly, although their existence was first proposed around the end of the eighteenth century. Back then, you'll recall, light was considered to be corpuscular so that it made sense to propose that the gravitational pull of very massive stars might prevent any light corpuscles produced on the star from being able to leave it. This would result in such massive stars appearing to be 'black' or invisible to observers on earth. When the wave theory of light gained ascendancy a hundred years later, the idea that such black stars may exist was discredited. Einstein's postulate that light was both a wave and a particle, revived this possibility once again."

"If they are actually invisible stars, then why are they called black *holes*?" Beth was prompted to ask.

"There is little doubt these days that there is a great deal of matter in the universe that is not directly visible to our telescopes on earth or in outer space." I replied. "Some of it is in the form of huge stars that have become so massive that their gravitational fields cause them to collapse into even smaller and denser masses. At some point, the field then becomes so strong that visible light can no longer escape from it and a black hole is born. The origin of the name is historical."

"So Einstein first demonstrated the equivalence of energy and mass. This enabled him to postulate that light photons could be a form of energy, expressed by a wave, or a mass, in the form of a particle." Beth was recapitulating for her own benefit. "And, finally, he showed how astronomical observations should be properly interpreted. Not bad for a kid who had trouble getting admitted into college on his first try."

# Chapter Thirteen:
# Breakfast of Farina

## *Does Anyone Believe in Quanta?*

"Today I challenge you to come up with a topic in physics that relates to the farina I cooked for breakfast." Beth announced with a twinkle in her eye.

"As long as you serve it with cinnamon sugar," I replied, "I shall accept your challenge."

"Let's see," I stalled while trying to think of an appropriate topic. "If you look at a bowl of farina from a distance, it appears to be a white continuum of some sort. But from a closer perspective, it is evident that the farina is made up of individual grains that are not only identical to each other but also to the entire porridge. The idea that a continuum of something could be represented alternatively by an assembly of individual units of that thing actually occurred for the first time to a German physicist in the year 1900."

"If you are talking about the atomic theory of matter, I thought it had been proposed much earlier than 1900." Beth stated confidently.

"I'm not talking about atoms," I responded, "whose individual properties are actually quite different from those of the matter that they form.

Instead, consider energy, a construct of physics that we discussed at one of our previous breakfasts. You will recall that we can't observe energy directly, but we know that it exists because we can measure the work that it produces. For example, we know that water flowing in a stream can rotate a paddle wheel which, in turn, can rotate an electric generator."

"And the electricity produced can charge a battery which can later be used to operate a motor boat." Beth joined in. "I remember all of these examples of energy transformation and conservation."

"Yes, once this concept of energy was finally accepted in the nineteenth century," I resumed, "it was natural to assume that energy varies in a continuous manner. For example, as one increases the speed of a moving object, its kinetic energy increases in proportion to the square of the speed.

What's more, we know that the speed and, therefore, the energy can be varied continuously rather than in a step-wise manner."

"But, if energy is just a convenient invention of physicists and one can't even see it," Beth couldn't restrain herself, "why would anyone want to claim that it consists of individual chunks the way that farina does?"

"To win a Nobel Prize?" I asked facetiously. "Picture a burner on an electric range or an iron bar placed over burning coals, as was the practice when we relied on horseshoes instead of rubber tires for our transportation. The bar first turns red and, as it gets ever hotter, it began to glow with white light."

"Is that where the expressions 'red hot' and 'white hot' originated?"

"Indubitably." I responded. "Actually, if you were to measure the intensity of the glow for each color or energy of the light given off, you would find that the intensity distribution shifts to higher frequencies as the energy of the emitted light color increases (Fig. 36). Physicists had observed these distributions well before the end of the last century, but theories proposed to explain their observations met with only limited success."

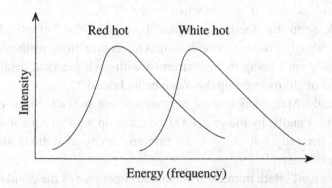

Fig. 36. Experimental emission curves obtained for a metal heated to two different temperatures. Note that the entire spectrum and its maximum are shifted to higher energies at the higher temperature.

"I should think that this is a problem ideally suited for the application of the laws of thermodynamics." Beth piped up.

"You are quite right," I happily agreed, "and here is where matters stood. Toward the end of the nineteenth century, it was generally believed that iron bars were made up of individual iron atoms. Even though no one knew very much about these atoms, it was reasonable to assume that they could absorb heat by increasing their respective kinetic energies and, subsequently, re-emit that energy in the form of visible light. As you correctly noted, such a model should lend itself to analysis by statistical thermodynamics since huge numbers of iron atoms are involved. In fact, several theories were developed but they did not match the experimentally observed emission spectrum completely."

"I sense a major departure from existing beliefs about energy coming up."

"Once more, your foresight is right on target. A former student of a highly regarded German professor of thermodynamics began to consider various theories that might better fit the observational data. After becoming a professor of physics himself, Max Planck undertook the task of matching the experimental emission curves with calculated ones. This required that he derive a continuous spectrum of all colors, whose maximum intensity shifts to shorter wavelengths (higher frequencies) as the radiating metal's temperature is increased (Fig. 36). What's more, the theory had to be independent of the kind of metal being studied since all metals emitted similar spectra when heated."

"Duck soup for thermodynamics, I should think," Beth observed condescendingly. "Since thermodynamics treats everything mathematically, it obviously isn't going to be concerned with such practical details as to what kind of atoms make up the metal being heated."

"Actually Max Planck tried various models but was only partially successful. Finally, by the year 1900, he came up with an equation that fit the data perfectly but failed to fit into any accepted notions about the behavior of matter."

"*Gevonnen!*" Beth mimicked the famous utterance of the drunken jailer in the third act of Strauss's operetta *Die Fledermaus* (uttered after winning a bet he had made with himself). "I did predict the coming of a revolution in ideas about energy."

"The revolution launched by Planck was much more profound than he even imagined. What Planck had discovered was that he could fit the experimental emission curves if he assumed that the atoms can absorb energy in discrete amounts only rather than continuously, as had been believed up till then in accord with classical theory."

"Point of information, please." Beth interrupted. "I understand the difference between a discrete set of beads in a necklace and a continuous string or rope. But how does this translate to the concept of energy?"

"Remember my comment about the farina that we are eating? When I pointed out that it may look to be continuous but is actually made up of discrete grains?" I asked. "Well compare the bowl of farina to a bowl of jello which is truly continuous and contains no identifiable discrete parts."

"I get it!" Beth observed happily. "An energy distribution like the emission curve from a hot iron bar (Fig. 36) appears to be continuous whereas it could actually be made up of discrete subunits of energy."

"Quite so," I was happy to agree. "Planck's postulate applied to the absorption of energy and it only worked, by the way, if the allowed quanta had energies that were multiples of a fundamental unit of energy that Planck called a *quantum* and designated by the letter symbol *h*. Why this postulate led to the correct fit of the observed emission curves puzzled Planck greatly. As a result, he didn't try very hard to find other applications for his new theory and left that up to others."

"I don't understand one thing," Beth now asked, "if the spectrum emitted by the heated metal consists of a continuum of light of all energies, why do the atoms absorb the heat in discrete quanta?"

"This is what also puzzled Max Planck." I replied. "Five years later, Albert Einstein proposed that the emitted light also consisted of discrete bundles of energy subsequently named *photons*. As Einstein showed in his Nobel Prize-winning explanation of the photoelectric effect, such photons act like particles whose energy, expressed in integer multiples of the same fundamental unit of energy determined by Planck's constant *h*, can knock electrons out of a metal. It is an amusing footnote in the history of physics that Einstein described these ideas a decade later to an audience that included Max Planck. At the end of it, Planck thanked Einstein for supporting his quantum theory but advised the *Herr Doctor* to be cautious in expanding it to include other physical phenomena."

"Hasn't the quantum theory found much wider applications than this?" Beth inquired. "Isn't the modern theory of the atom also based on it, for example?"

"Quite right! In fact, just as Einstein's theories of relativity subsumed Newton's classical theories of mechanical motion and gravitational attraction, the quantum theory has been expanded to include physical phenomena throughout the universe."

"I'm still puzzled by the fact that we observe light as a continuous range of colors or energies and, yet, you say that it consists of discrete quanta of energies called photons." Beth persisted. "Has anyone actually seen individual photons?"

"We are now entering a realm of physics where direct observation becomes well nigh impossible." I replied. "We shall have to rely increasingly on indirect evidence to support our theories explaining natural phenomena."

"That sounds like circumstantial evidence to me." Beth observed. "Is it as reliable as direct observation?"

"I hope that, by the time we conclude our breakfast discussions of physics, you will share my confidence in the validity of modern physics." I tried to sound reassuring. "For example, ninety years after Planck's original postulate, an artificial satellite was launched specifically to explore the radiation that fills all of outer space. The data it collected fits the original equation derived by Planck perfectly!"

"I wondered whether that would have sufficed to reassure poor Max that his theory was for real?"

"I suspect that the Nobel Prize that he was awarded in 1918 may have done that. By now, of course, the quantum theory has become so fundamental in physics that no serious physicist questions it any longer."

## What About Atoms?

"You began our discussion this morning by likening the farina to an energy continuum made up of discrete subunits of energy called quanta." Beth tried to draw her own word picture. "When a collection of metal atoms is energized to white heat, it absorbs the thermal energy in multiples of these discrete quanta, according to Planck, to produce a continuum of visible radiation. According to Einstein, the energy of the light emitted is also quantized. Am I right so far?"

"You're doing very well!"

"Now then, what about an atom?" Beth asked next. "Since an atom is the basic unit from which matter is made up, is it correct to think of it as a quantum of matter?"

"We should be careful in our choice of words so as to preserve their correct meanings." I responded. "In physics, the term quantum is limited to the definition given it by Max Planck, namely, as the smallest unit into which all energy can be divided. He called it a quantity of action or a quantum, for short. Matter, like the delicious farina that you served me this morning, may be composed of identical units of farina grains, just like an iron bar is composed of identical iron atoms, but physicists would not call them quanta."

"Well how does an iron atom differ from a piece of iron metal?" Beth asked provocatively. "Other than that an atom is much smaller."

"Let's compare just one property of iron," I gladly began to explain. "We saw that we could heat an iron bar until it emitted white light. If you pass that light through a narrow slit and then through a glass prism, what you'll see is that it is made up of a rainbow of colors. The same is true of the light from any incandescent lamp whose filament is made up primarily of tungsten atoms."

"As Newton demonstrated three-hundred years ago," Beth confidently announced, "this proves that the white light emitted by a collection of metal atoms is made up of a continuous spectrum of colored lights each having a characteristic frequency."

"Right on!" I glowed with pleasure. "Even though iron atoms are quite different from tungsten atoms, their collective behavior when they are heated is very similar and in full accord with the quantum model proposed by Planck."

"But suppose next," I continued, "that we could separate the metal atoms from each other, as happens when we vaporize the metal. Individual metal atoms still can absorb additional energy in a gaseous state. When they subsequently radiate this extra energy, however, what comes off is not a continuous spectrum but a few discrete photon energies that are characteristic of the radiating atom."

"Oh I think that's what our daughter did in her spectroscopy laboratory in college!" Beth exclaimed. "That's how chemists and other scientists identify the atomic constituents when they wish to chemically analyze matter."

"Another example of the practical application of physics," I concurred with pride. "Actually, a chemist and a physicist at the University of Heidelberg combined their talents in the middle of the nineteenth century to demonstrate that individual atoms emitted or absorbed light whose energies were as unique as the fingerprints of individual people. Using a simple prism, they were able to analyze the atomic compositions of the sun and other distant stars for the first time ever. In this way (Fig. 37), they also discovered the existence of a number of elements that had not been previously identified on earth."

"Didn't the English chemist, John Dalton, originate the modern atomic theory of matter at the start of the nineteenth century?" Beth inquired.

"Dalton might be classified more appropriately as a meteorologist nowadays." I observed. "While studying the way gases and liquids mixed with each other, Dalton concluded that the ultimate particles making up matter must combine in proportions that could be expressed by simple numerical ratios. In this way, he gave credence to the purely speculative ideas of the ancient Greeks, notably one called Democritus, who had posited over two-thousand years earlier that all matter was composed of invisible and indivisible particles that he named atoms."

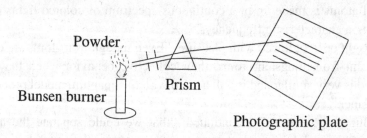

Fig. 37.   The chemist, Robert Bunsen, and the physicist, Gustav Kirchhoff, used the hot flame of a Bunsen Burner to excite atoms making up various compounds. After passing through a slit defining a narrow beam, the emitted light was refracted by a prism at an angle characteristic of its energy. By recording the emitted spectral lines in such a spectrometer, it is possible to identify the emitting element from its 'fingerprint'.

"Coming back to atomic spectra," Beth disapproved of my digressions, "how did Planck account for the discrete spectra of individual atoms as compared to the continuous spectra emitted by hot metals?"

"He didn't." I replied. "In fact, this was just one of a growing list of observations that physicists were unable to explain at the dawn of the twentieth century."

"Well what, exactly, *did* physicists know about atoms a hundred years ago?" Beth persisted.

"To begin, there was a growing body of chemical knowledge that clearly pointed to the existence of unique atoms from which all matter was made." I explained. "Also, J. J. Thomson, the discoverer of the electron, suggested that individual atoms were like tiny globs of a positive jelly within which individual negative electrons were embedded. The jelly accounted for the mass of an atom while the electrons could gain energy by vibrating within this jelly. In this way, he could explain why atoms could absorb energy and later re-emit it. What Thomson did not know was how to prove that this was a correct model for an atom."

"How would you like some more farina?" Beth interposed. "With some strawberry jelly in it?"

"Yes, I would thanks. But I prefer strawberry preserves to jelly, thank you."

"Returning to your tale about the atom," Beth observed as she ladled more cereal. "I sense some dramatic breakthrough in the offing!"

"Actually, science, in general, and atomic physics, in particular, does indeed move forward by leaps and bounds."

"Don't tease," Beth insisted. "What happened next?"

"By 1895, Thomson had been joined by a young physics student from New Zealand, destined to succeed Thomson as Cavendish Professor at Oxford University and to be named First Baron of Nelson and Cambridge in 1931."

"I know that the British Crown awards knighthoods to distinguished individuals," Beth observed, "but is it usual for scientists in England to receive baronets?"

"Highly unusual," I responded, "but then Ernest Rutherford was a highly unusual physicist. In 1908 he was awarded a Nobel Prize for the discovery that certain *alpha rays* emitted by radioactive minerals were actually positively charged helium atoms. Two years later, he decided to bombard gold foil with these alpha rays to study how they would interact with matter."

"When you speak of bombarding with alpha rays," Beth interrupted, "is that at all like spraying the gold foil with helium atoms out of some kind of spray can?"

"Not a spray can, exactly, because the alpha particles *are* sent out by the radioactive atoms in all directions at very high speeds."

"If they fly out in all directions, like the seeds of a ripe dandelion, how does one get them to bombard a particular metal foil?" Beth wondered out loud.

"Oh, I know!" She answered her own question. "By placing parallel slits in front of the foil, only a directed beam of alpha particles can reach the foil."

"Yes, by limiting the incoming beam of particles," I nodded, happily resuming my story, "Rutherford's assistants could measure quite accurately the directions in which the alpha particles were scattered by the gold foils. The only way Rutherford could explain the resulting observations was to assume that each gold atom was composed of a tiny positive nucleus surrounded by negative electrons. The volume of this nucleus, however, had to be ten-thousand times smaller than the estimated size of the entire gold atom. To appreciate these relative magnitudes, think of an atomic nucleus as a single pea. The entire atom then would have a radius equal to the length of three football fields!"

"What then fills in the huge spaces between nuclei of adjacent atoms in a metal?" Beth asked in wonderment.

"Several models were put forward in which the negative electrons balancing the positive charge on the nucleus were assumed to circle the nucleus in appropriately large orbits."

"You mean the same way that the planets circle around the sun?"

"Very similarly," I replied. "In fact, such a planetary model seemed highly probable except for two very serious problems."

"I bet one of them had to do with how to keep the negatively charged electrons from being drawn into the positively charged nuclues." Beth volunteered.

"That's very perceptive. This was exactly why Thomson had postulated smeared out positive jelly for the structure of an atom."

"Wait a minute." Beth couldn't hold back. "The planets circling the sun are attracted to it by gravitational forces, yet they continue in their orbits. Isn't that due to a balance between their linear inertia, which would keep them moving in a straight line, and the centripetal force due to the sun's gravitational attraction?"

"Terrific!" I couldn't contain my joy. "You are quite right to suppose that a centripetal electrostatic force could serve to overcome an orbiting electron's linear momentum, in full agreement with Newton's law of motion."

"Just like the gravitational attraction of the sun," Beth resumed, "the centripetal electrostatic attraction of the nucleus produces a change in the electron's direction of motion so that it follows a circular path around the central nucleus. That takes care of the first objection. So what was the second objection?"

"Actually it comes directly out of the very clever way in which you described the resolution of the first objection." I replied. "According to Maxwell's equations, a charged object undergoing an acceleration must radiate energy."

"What acceleration?" Beth interrupted. "Why can't the electron follow a regular circular orbit at a constant velocity?"

"Got you this time!" I couldn't help teasing Beth. "Recall that any force produces an acceleration, even if it's only to change the direction of motion at an otherwise constant speed."

"You're right. The centripetal acceleration results from the centripetal force that produces the nonlinear orbit in the first place." Beth admitted ruefully. "But, if the orbiting electron radiates any of its energy, will not its radius decrease until it spirals into the nucleus?"

"And that's the second problem with the planetary model as it existed in 1912."

"So tell me, just what did happen in 1912?" Beth teased with a twinkle in her eye.

"A twenty-seven year old Danish physicist, Niels Bohr, left Thomson's laboratory in Cambridge to join Rutherford at the University of Manchester." I resumed. "In Rutherford's laboratory, Bohr saw the evidence for a planetary atom as well as its shortcomings. Consider the simplest atom, hydrogen. It consists of a positive nucleus of charge $+e$ around which a single electron of charge $-e$ circles. Yet even this very simple atom can emit an entire spectrum of characteristic lines when it is suitably energized. This means that the circling electron can have more than one possible energy! And all this, in total disagreement with Maxwell's elegant theory."

"That's a fascinating dilemma!" Beth exclaimed "Bohr knew that the experimental evidence could not be wrong and, yet, there seemed to be nothing wrong with prevailing theory except, of course, that it was unable to explain the experimental evidence!"

"Knowing all this, Bohr proceeded to develop a model for the hydrogen atom that was based on two simple postulates." I continued. "First, he said, let's assume that the one electron in hydrogen can revolve around the nucleus in one of several orbits without radiating energy!"

"You mean that there are several orbits that the sole electron can travel in?" Beth sounded puzzled. "How does it get from one to another?"

"We'll come to that in a little while." I replied. "But, yes, there are several allowed orbits in which the single electron of hydrogen can circle its nucleus (Fig. 38) without radiating any energy. You understand that he

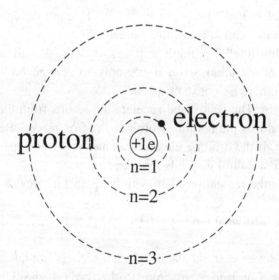

Fig. 38.   Simplified Bohr model for hydrogen. (Not to scale as the atom's radius is 10,000 times larger than that of the nucleus, which is called a *proton*.) The one electron of hydrogen can occupy any one of the orbits indicated and identified by their principal quantum number n.

had no physical evidence that this was actually the case, so that this really was a working hypothesis when first proposed. Nevertheless, Bohr numbered these orbits 1, 2, 3, ... and named this integer the *principal quantum number n*. Because the orbits associated with these values of *n* were the same for all hydrogen atoms and because they obviously were stable orbits, Bohr called them *stationary* orbits.

The second postulate that Bohr proposed deals with the *angular momentum* of the orbiting electron. Any object moving in a curve has an angular momentum that keeps it moving along that curved path in the same way that the linear momentum of an object maintained its motion in a straight line."

"You mean its momentum is responsible for its inertia?" Beth interjected.

"Very good!" I agreed. "The angular momentum *is* directly related to the inertia. Here again, Bohr made a very arbitrary decision that the angular momentum of an electron in each circular orbit must be a multiple of Planck's constant *h* divided by $2\pi$, where $\pi$ (= 3.147 ...) is the Greek letter *pi*. You remember from geometry that *pi* figures prominently in calculations involving curvatures of all kinds. Bohr chose integers having positive values ranging from zero upwards as possible multipliers and this became his *second quantum number*."

"What made Bohr fix these magnitudes in terms of Planck's constant?"

"Several factors guided Bohr's thinking," I responded. "You may recall that Coulomb had shown that the energy of a charged particle was inversely related to its separation from another charged particle. The further apart they are, the lower their interaction energy. The same, therefore, should be true of the negative electron circling a positive nucleus. Bohr assumed that the radii of the allowed orbits in hydrogen increased as the square of the principal quantum numebr, *n*. It then followed that the electron's energy would decrease in successive orbits as *n* squared increased.

If you wonder how Bohr came to make all these assumptions, you should know that this fit very nicely a discovery that a Swiss high-school teacher, named Johann Jakob Balmer, had made some twenty-five years earlier. Balmer had deduced that the frequencies of experimentally observed spectral lines of an excited hydrogen atom could be related to each other by a simple formula involving ratios of small integers squared. Bohr also knew about Einstein's discovery that the photon energies involved in the photoelectric effect were equal to Planck's constant multiplied by the photon's frequency, and that Planck, himself, had postulated that energy could be absorbed only in integer multiples of $h$ times frequency."

"All this sounds good and seems logical," Beth now observed, "but that doesn't mean that I understand what Bohr did or why he did it."

"Well, let me try another approach." I was thinking furiously. "Let's grant Bohr the right to postulate his stationary or fixed orbits. Let's also accept that their radii increase as the square of the quantum number $n$ assigned to them (Fig. 38) increases. If we also accept Planck's idea of energy changing only in multiples of a basic quantum proportional to $h$, and Einstein's conclusion that light photons have energies that are identical multiples of $h$, then Bohr's model for hydrogen fits exactly the observed line spectra in accord with the numerical formula discovered by Balmer in 1885."

"How does it do that?"

"First of all, the one electron of hydrogen has to occupy one of the stationary orbits prescribed by Bohr. The energy of the electron is characteristic of that orbit so that the specific energies associated with each orbit are the only energies that the hydrogen electron can have. Thus, to gain or lose energy, the hydrogen electron must go from one allowed orbit to another. This means that it can gain or lose energy in amounts equal to the difference between the energies of the allowed orbits and no others!"

"And that's how allowed and forbidden energy values arise in atoms!" Beth exclaimed triumphantly.

"You've got it!" I was equally delighted.

"And that's why the quantum theory underlies all the modern physics."

"You're right again!"

"Now, if only I can remember what I just said next time we sit down to one of our extended breakfasts!"

# Chapter Fourteen:
# Breakfast of Danish Pastry

## *What's a Wave Function?*

"I have been inspired by our last discussion of Bohr's atomic model to buy some Danish pastry at the new bakery that opened in town last week." Beth announced the following morning.

"That's very naughty of you because you know how I love this flaky pastry and what it does to my not-so-successful attempts to control my diet." I responded. "But what more can I tell you about the Bohr model?"

"I've heard and read enough about Niels Bohr to know that his atomic model was a major advance in the development of modern physics." Beth observed. "But, from what you told me so far, it seems to have been based on some arbitrarily selected assumptions. How well did it actually fit experimental data on atoms?"

"One could say that Bohr's assumptions were 'forced' on him by the available experimental data in the same way that Planck's postulate of quantized energies was totally arbitrary but made it possible for him to match theory to experiment exactly." I responded in part. "As to fitting existing data, recall that the Bohr model was devised for hydrogen having only one electron. When it was extended to other atoms, additional postulates had to be made. They became known as quantum selection rules and involved two additional sets of integers which are also called quantum numbers. All in all, things were getting more and more convoluted."

"Let me make sure that I've got it right about those early developments." Beth recapitulated: "Planck kicked off the quantum theory by matching the observed spectral data to thermodynamical calculations. He did this by arbitrarily postulating that energies have to be quantized. There was no physical basis for what he did except that it somehow 'fit' the experimental curves. Am I right so far?"

"Absolutely."

"Not to be outdone, Bohr made some equally arbitrary postulates about the existence of certain stationary orbits to which electrons in hydrogen atoms are confined. Not only was there no physical basis for these assumptions, but they flew in the face of accepted theories, in particular, the relatively recent and revolutionary equations of Maxwell. How am I doing so far?"

"You're right on!"

"Caught up in this euphoria and unmindful of the world war and its aftermath in Europe, physicists went merrily forward adapting these unfounded assumptions to fitting ever more experimental data about atoms of all kinds. Didn't anybody stop to wonder: hey, this may be fun but is it really physics?"

"You're quite right." I was glad for the opportunity to elaborate. "By the 1920's, it had become clear to several physicists that a new philosophical basis was necessary to provide a rational and self-consistent conceptual framework. While these theoreticians were toiling in Germany, totally unexpected help was provided by a graduate student at the University of Paris. The son of a French duke, Louis de Broglie was searching for a common thread between Einstein's postulate of the wave-particle duality of light photons and Bohr's model for an atom."

"If he was the son of a duke, didn't that make him a prince?" Beth's interest rekindled. "I like the idea of a French prince riding to the rescue!"

"As happens in any good fairy tale, our charming prince first had to overcome the pall cast on him by his skeptical professor." I resumed. "Was it possible for an electron, that Thompson had shown to be a negatively charged particle, to also behave like a wave, an idea that Hertz had actually derived 'erroneously' from some of his early experiments with cathode-ray tubes? If, however, one were to accept such an unorthodox premise, de Broglie argued, then a number of puzzling aspects of the Bohr model would actually make physical sense!"

"If I recall correctly, a light photon has no mass." Beth wanted to be sure. "According to Einstein, it is a bundle of energy that can act like a particle or a wave. Even if French professors in 1920 could swallow the notion of a wave-like but massless 'particle,' I can imagine how they must have bridled at the notion of a real particle that has a known mass also acting like a wave. How did de Broglie overcome their skepticism?"

"Like this: suppose the circumference of Bohr's stationary orbits was exactly equal to *n* times the wavelength of an electron. (Where *n* is the quantum number assigned to a particular orbit.)" I continued. "Then the electron wave would just fit into these orbits (Fig. 39) and explain their stability as well as why the quantum number *n* had to be an integer."

"I wonder how de Broglie knew what the wavelength of an electron should be?" Beth speculated as she clung to her image of a French prince riding to the rescue.

"Influenced by Einstein's equation relating energy to mass and Planck's definition of the energy of a quantum, de Broglie postulated a comparably simple equation relating the momentum of an electron to its wavelength. Although derived by the application of pure logic, a process of reasoning French scientists seem to favor, de Broglie's postulate was every bit as bold a leap of his imagination as had been Planck's and Bohr's before him."

"I'm still curious how he 'sold' this idea to his professor in Paris." Beth persisted.

"Not easily," I assured her. "It was only after Einstein had read and approved de Broglie's thesis that the University of Paris agreed to grant

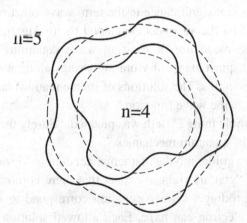

Fig. 39.   Two of Bohr's stationary orbits (circles) for hydrogen showing how de Broglie's electron waves fit into them. (Only the orbits for *n* = 4 and 5 are illustrated.)

him the doctoral degree. Five years later, the Norwegian Academy awarded him the Nobel Prize in physics for the same discovery."

"Other than fitting the Bohr model for an atom, is there any independent proof that electrons have wave-like properties?" Beth asked. "And how did de Broglie's postulate help the theoreticians looking for a sound conceptual basis for a quantum theory that had been based, so far, only on a series of *ad hoc* postulates?"

"As often happens in the evolution of science, a couple of physicists studying vacuum tubes at the Bell Laboratories in the United States actually had observed the wave nature of electrons, but failed to recognize it as such. Their observation of electron scattering by metal components in the tubes at certain angles only, was cited subsequently by a couple of German physicists as proof that electrons could be diffracted in the same way that waves are."

"Who got the Nobel Prize?" Beth asked mischievously. "The Americans or the Germans?"

"Actually only the senior American investigator did, but then he had to conduct some further experiments to confirm and amplify this early finding." I replied. "More importantly, this discovery helped direct the thinking of several German physicists who were trying to create a new theoretical structure at that time. The theory that resulted is currently known as *quantum mechanics* and makes use of a *wave function* to describe the electron."

"Is there a special significance to the term wave function?"

"One of the first theories was developed by an Austrian called Erwin Schrödinger. He made use of existing wave equations developed for describing the displacements of vibrating strings along with de Broglie's postulated wavelengths. The solutions of such equations have a wave-like form, hence the name wave function."

"Is that the whole thing?" Beth was puzzled. "Surely there was more to the new theory of quantum mechanics."

"Even without going into the mathematical details," I tried not to sound condescending, "you are right to expect that there is more. The possible solutions of Schrödinger's wave equation correspond to specific energy values that the electron can have. Each allowed solution contains three integers which turn out, not surprisingly, to be the very same quantum numbers that Bohr and his colleagues had postulated previously to extend the hydrogen model to other atoms. Schrödinger's equation, however, provided a sound theoretical basis for applying these quantum numbers to explain the observed spectra of all the atoms, including optical spectra and x-ray spectra. At last it was possible to provide a physical basis for the periodic table of elements that the Russian chemist, Dmitri Mendeleyev, proposed toward the end of the nineteenth century."

"It begins to sound like a jigsaw puzzle fitting together." Beth observed. "By the way, did I correctly understand you to say that these explanations were based on the three quantum numbers associated with each allowed solution of Schrödinger equation? What, then, does the wave function tell us about the electron?"

"You have this tendency to ask two separate questions in the same breath." I grumbled. "To be precise, a fourth quantum number is necessary to account for all known aspects of atomic spectra and atomic properties. This need was determined empirically in the U.S. and deduced theoretically in Germany at about the same time.

As to the wave function," I continued, "it tells us where an electron is located but in a fuzzy way. By this I mean that the wave function of any electron extends throughout all of space. It has maximum values at exactly where the original Bohr orbits were postulated to exist, but it is not limited in space to these radial values only."

"Hold on a minute!" Beth sounded agitated. "Is this some kind of new hocus-pocus? Now you see it and now you don't? How can the wave function describing the location of an electron place it in a stationary orbit and yet have a nonzero value elsewhere in space?"

"It so happens that this is as good a description of nature as we need. The problem arises when we try to verify where the electron actually is." I tried to sound reasonable. "We can observe the displacement of a rope just as it is predicted by the wave equation. To do this, we shine a beam of light on it and the light photons scattered back to our eyes tell us where it is at any moment. Should we direct even a single photon at an electron, however, it will necessarily interact with that electron in a way that will change the electron's location, or motion, or energy. Even if we had some way to observe the photon after its interaction with the electron, the information we would receive would no longer correctly describe that electron."

"So we have another dilemma." Beth observed somewhat smugly. "We can calculate what electrons do but we can't actually observe them doing it. How do physicists make sense of this kind of confusion?"

"Please remember that physics tries to describe the nature of the physical universe we inhabit. It does not have the ability to modify the universe so as to make it fit some theoretical construct, no matter how elegant such a construct may be." I began to sound somewhat ponderous even to my own ears. "At the atomic level, we are finding that we must accept a universe that behaves quite differently from the one that Newton described so aptly nearly 400 years ago."

## Born and Heisenberg Have the Answer

"Even though I don't fully understand it," Beth announced, "I am really intrigued by the new world of quantum physics. Can you tell me more about the mysterious nature of the electrons?"

"Schrödinger was aware of the problem of reconciling his wave-equation solutions with an electron that he knew was a real particle." I resumed my narrative. "Keep in mind that the wave nature of real particles like electrons was mystifying not only de Broglie's professors in Paris, but all scientists throughout the world, including Erwin Schrödinger. A conceptual breakthrough came soon after Schrödinger published his theory. Professor Max Born at Heidelberg University, demonstrated that, if you square the wave function at any point in space, what you get is the probability that the electron is actually present at that particular place. Called the *probability density*, the wave-function squared tells us what the probability is of finding the electron at any particular location."

"What you are now saying is that the square of a wave-function solution of Schrödinger's equation will have its maximum value wherever the wave function does." Beth was making sure of her correct understanding. "This must occur at one of the radii of Bohr's stationary orbits. But, since the wave function does not drop to zero away from this specific value, there is still a small but finite probability remaining that the electron may be somewhere else."

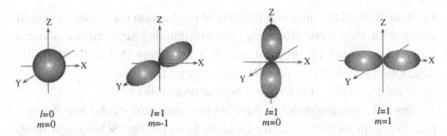

Fig. 40.   Some of the possible *orbitals* that electrons can occupy in an atom. The shaded regions represent the most likely locations (probability density) for the electrons near the nucleus at the center. The wave functions used correspond to solutions involving different quantum numbers $l$ and m.

"That's a very good description of the picture that Born's analysis provides. We now talk of an *orbital* (Fig. 40) rather than an orbit for an electron. Since we can't actually observe an electron directly anyway, a probability density is a most useful quantity. What we *can* do is repeat a measurement involving an electron a large number of times and, when we do, the probabilities calculated using wave functions match our observations exactly!"

"Didn't thermodynamics already deal with large numbers of quantities rather than individual instances?" Beth wondered. "How does quantum mechanics differ from this?"

"What Born's probability density does is predict the likeliest place where an electron *may be*. In doing this, it prevents us from knowing where the electron actually *is*.

Put another way," I continued, "classical mechanics gave us a *deterministic* view of the world. Quantum mechanics, conversely, gives us a *probabilistic* view instead. According to Newton, if you know the cause of an event, you can predict the outcome. According to Born, you can only predict how likely that outcome will be."

"How probable is it," Beth teased, "that with a bit of luck someone will come up with a more precise theory allowing us to be more deterministic about electrons?"

"Schrödinger, himself, wondered whether a more precise theory might be possible." I replied. "And I'm sure that you have heard the assertion attributed to Einstein that 'God is not a crap shooter'. In particular,

Einstein disliked the quantum-theoretical implication that observers could affect what they were observing. Notwithstanding such unease about a universe governed by probabilities, it is not likely that a deterministic description of subatomic particles will ever be feasible."

"Would you like to place a bet on that probability?"

"Several other theoreticians have derived quantum-mechanical theories based on approaches that differ radically from the one Schrödinger used. Their results agree remarkably well with each other and with the original predictions of the wave equation even though the later theories are considerably more comprehensive. Developed by Werner Karl Heisenberg, one of these showed that an actual limit exists on how precisely it is possible to make any physical measurement. Popularly known as the *uncertainty principle*, this result states that there is a lower limit on how precisely we can measure any two complimentary quantities. This lower limit, by the way, is nothing but the ever-present Planck's constant divided by $2\pi$!"

"I've heard the Heisenberg uncertainty principle mentioned before, but I never understood how it applies to actual observations." Beth noted.

"Take any two complimentary quantities, such as the energy of an electron at some specific time, or its momentum or speed at some given point in space. The Heisenberg principle states that, the result of multiplying the experimental uncertainties in measuring two such quantities must always be larger than $h$, Planck's constant."

"Oh, I see," Beth exclaimed, "If you decrease the uncertainty of one of the pair, you must increase the other accordingly, since their product can't be less than $h$!"

"Well said!" I beamed. "If, for example, we take the very best measured value that we have been able to determine for the mass of an electron and ask what will be the smallest error in measuring its location after assuming a most reasonable magnitude for the electron's speed, we shall find that the uncertainty in locating the electron is as large as the dimension of an atom! In other words, it is the nature of nature to keep our picture of an atom fuzzy forever."

"Before we leave this high-flown discussion of the whereabouts of an electron," Beth asked, "is there any reasonable way to reconcile the dual nature of light photons and particles of matter like electron?"

"You mean, does Einstein's assumption of a wave — particle duality for photons match the duality posited by de Broglie for electrons?" I rephrased Beth's question.

"Yes."

"The *photoelectric* effect shows that a photon can act like a particle in knocking out an electron from a metal. We also know that the diffraction of light requires that photons must behave like waves. So we accept the dual nature of photons. Analogously, we have no trouble accepting the fact that an electron is a negatively charged particle with a known mass. Once we prove that electrons can be diffracted, then we should have no problem believing that they can act like waves. The existence of electron microscopes is based on the ability of electrons to behave like waves."

"I accept all that," Beth puzzled. "But I still have difficulty in picturing a particle that has no mass at all or a more massive particle that can act like a wave. Do you or I have waves associated with us?"

"Yes we do, except that our wavelengths are so short that they are not perceptive." I explained. "In order to convince you of the reality of all this, let me tell you about Arthur Compton's key experiment. While a graduate student in St. Louis, Compton was studying the scattering of x rays, which are also a form of electromagnetic radiation. He observed that x rays can be scattered by matter at predictable angles without any change in their wavelength, in full agreement with Maxwell's classical theory. What was totally unexpected was that the same x rays were also scattered at other angles after a change in their wavelength took place (Fig. 41). Nothing in classical theories developed to date could explain this observation."

"You mean at the time there was no way to explain why some of the scattered x rays should have a different wavelength while others did not?"

"Exactly!" I resumed. "Nevertheless, Compton persisted and, some ten years later, in 1923, he proposed a remarkable explanation: the x rays whose wavelength remained the same were obviously being scattered according to Maxwell's laws governing the scattering of electromagnetic waves. By treating x-ray photons and electrons as if they could behave like billiard balls, Compton argued that x-ray photons should be able to bounce off electrons with a change in both their energies and momenta as predicted by the classical mechanics of Newton. So x rays can act like waves and like particles in the same experiment and all this in accord with

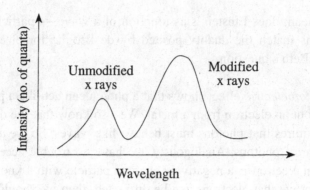

Fig. 41.  X rays are scattered by a block of carbon at two angles. The x-ray photons giving rise to the smaller peak have the same energy as the incident x rays while those making up the second peak have had their energy modified in accord with Compton's calculations.

classical theories! To believe this calculation, all you have to do is accept the fact that photons and electrons can carry on as if they were hard impenetrable particles!"

"1923 was right around the time when the various quantum theories were being developed." Beth now observed. "Were any of them able to account for this unexpected results?"

"Yes they were and most satisfactorily, I might add. Nevertheless, I shall leave you to ponder this puzzle: in Compton's experiment, an x-ray beam striking some substance like carbon is scattered at two angles (Fig. 41). At one angle, the scattering is explained by assuming that x rays are waves. At the other angle, however, we must assume that they act like particles. We seem to be able to change the character of the x rays just by changing the angle at which we set our detector!"

"You, yourself, said that physicists are mere observers of nature." Beth did not take long to rise to my challenge. "Obviously, photons are both waves and particles at the same time! Einstein was right. And, I suppose, so was de Broglie in predicting that particles of matter also were wave-like in their behavior. I just hope that all these clever physicists were appreciated for their courageous effort."

"Fortunately they were. They and others working in the same arena eventually received Nobel Prizes."

# Chapter Fifteen:
# Breakfast of Waffles

## *Atoms Can Be Fun*

"Before you smother your waffle with strawberries and whipped cream, look how it is divided into large squares which, in turn, are divided into smaller squares in a most regular manner." Beth announced as she placed a steaming waffle in front of me on the breakfast table. "Didn't ancient Greek scholars believe that all matter ultimately could be subdivided into like but invisible entities that they named *atomos*?"

"There were actually two competing schools of thought in ancient Greece." I replied while pouring hot maple syrup on top of the resplendent waffle. "One group, to which Aristotle belonged, held that matter was continuous. The other, headed by Democritus, contended that 'there are only atoms and empty space — all else is mere opinion'."

"Which only proves that arrogance was passed down to modern scientists along with the atoms!" Beth was in good form this morning.

"Democritus had many interesting ideas about atoms." I continued savoring my waffle. "In pondering why iron pots themselves were rigid while the water inside a pot was liquid, he speculated that iron atoms must have hooks on them while water atoms were slippery by nature. He also believed that the total number of atoms was probably infinite and that they must be in constant motion. These ideas were later expanded by Epicurus, who taught that atoms had a finite size rather than coming in a range of sizes. If atoms did vary in size, he reasoned, then some should be large enough to be seen."

"Wasn't Epicurus also the Greek scholar who opened a school in Athens to which he admitted women — a scandalous idea in 306 B.C.?"

Beth asked. "But among more contemporary scientists, was it chemists or physicists who picked up on the atomic model of matter?"

"Keep in mind that the current specialization among scientists evolved sometime in the nineteenth century." I responded. "Before then, scientists did not hesitate to address any questions about nature that piqued their interest. For example, the early gas laws developed by Robert Boyle and others depended on the existence of individual gas atoms. I believe Boyle is most often called a physicist, even though this term was unknown in his day. Similarly, John Dalton, who was trying to understand weather patterns, discovered the law of multiple proportions that states that any two elements always combine to form compounds in the same weight ratios. Modern chemists claim Dalton as the father of the atomist chemistry."

"What is the law of multiple proportions?" Beth wanted further elaboration.

"When analyzing two different gases known to contain only oxygen and carbon," I explained, "Dalton noted that the weight ratio in one case was 28% carbon plus 72% oxygen, and in the other 44% carbon plus 56% oxygen. The ratios of these two percentages are 2.57 and 1.28. If 1.28 correctly expresses the weight ratio of one oxygen atom to one carbon atom, then a ratio of 2.57 implies the existence of two oxygen atoms for each carbon atom in the other gas. By repeating this kind of analyses for numerous other gases, Dalton was able to show that atoms combine to form compounds in ratio expressible by small whole numbers."

"And that supports the notion that matter should be composed of individual atoms which combine in various integer ratios to form all the compounds on earth." Beth was pleased with her discovery.

"An unexpected boost was given to the atomic theory of matter by an English botanist. In 1827, Robert Brown described microscopic observations in which pollen particles suspended in a liquid executed erratic and continuous zigzag motions. Such *Brownian motion* can be understood if one believes that a liquid is made up of individual atoms or molecules that are constantly moving about at different speeds. A formal theory describing the erratic motions that such bombarding of suspended dust or pollen particles would produce was published by Einstein in 1905."

"That should have put the atomic theory on a firm footing." Beth volunteered.

"Actually, the unequivocal acceptance of the atomic theory of matter had to await the development of the quantum theory and Bohr's model of the hydrogen atom."

"Well would you agree that the early contributions to the atomic theory were more interested in how it affected chemical reactions than in the physics of what atoms were?" Beth persisted. "For example, didn't you tell me that Mendeleyev, a professor of chemistry at the University of St. Petersburg, published a grouping of the elements into regular periods that was based on their relative weights? But how could Mendeleyev weigh an atom?"

"He couldn't." I explained. "What Mendeleyev and his fellow chemists could do was determine the weight ratios of two elements like hydrogen and oxygen making up water. Once they established that their weight ratio was two to one, they were stuck. But then chemists had the brilliant idea to assign a weight of sixteen to oxygen which made the relative weight of hydrogen, the lightest element known, turn out to be very nearly equal to one."

"One what?" Beth asked. "One pound, one ounce, or one gram?"

"Neither. These numbers represent something called an *atomic mass unit* or *amu*, for short." I responded. "Once oxygen's weight is fixed at 16 amu, all the other elements making up the periodic table are assigned proportionate weights also expressed in amu."

"What then guided Mendeleyev in deciding how to group the elements in his table?"

"The way that they tended to combine with each other to form compounds." I tried to recall the explanation from my meager knowledge of chemistry. "For example, the elements in each column of the table tended to combine in similar ratios with the elements in the other columns."

"Despite its ability to account for similarities exhibited by the atoms appearing in the same column in Mendeleyev's table (Fig. 42), his groupings did not provide an explanation of why the atoms should fall into the proposed pattern." Beth persisted. "So I don't see that it had much practical value."

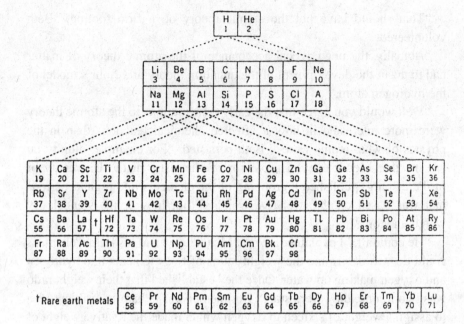

Fig. 42. Periodic Table of the elements. The vertical columns group atoms having like electronic structures and, hence, like properties. The atomic number of each element appears below its chemical symbol.

"Your view was shared by others of that time." I responded. "But let me show you some of the ways that the table proved its value. The Periodic Table originally contained some gaps which Mendeleyev claimed were atoms not yet discovered. He was mocked by his colleagues for this but, within fifteen years, the discovery of the element *gallium* (Ga) and *germanium* (Ge) silenced the scoffers and confirmed the general correctness of Mendeleyev's table. But probably the most significant aspect of the Periodic Table is the numerical progression of the elements which we now call their atomic number, Z."

"You mean Z = 1 for hydrogen (H), 2 for helium (He), 3 for lithium (Li), and so on? Isn't that just like a numerical sequence?" Beth now asked. "I guess I don't see any special significance in these numbers except that they increase in a regular manner."

"The true significance of this sequence of numbers remained obscure for some time." I explained. "Even the quantum-mechanical theories of

Schrödinger, Heisenberg, and Dirac couldn't, by themselves, provide a rational basis for the periodic classifications. It was left to Wolfgang Pauli, an Austrian theoretical physicist, to show that an additional restriction had to be placed on the possible energy values assigned to the electrons making up each atom. Called the *Pauli exclusion principle*, it finally provided an acceptable explanation of the Periodic Table of the elements. This also, by the way, opened up the whole field of quantum chemistry by providing the key to understanding how atoms can combine with each other."

"How difficult is it to understand these quantum theories?" Beth wanted to know.

"It's really not at all difficult, but it does require learning a few simple rules." I replied. "After that, it becomes a kind of game, like a crossword puzzle but much simpler. In fact, you may find it enjoyable, particularly once you understand how the patterns in the Periodic Table evolve and how and why specific atomic combinations are most likely to occur in nature."

"I'm game." Beth announced. "What are the rules?"

"Remember my telling you that solutions of the quantum-mechanical equations involved sets of integers called quantum numbers?" I asked. "Well, these numbers have interdependent values according to the following scheme:

The *principal* quantum number $n = 1, 2, 3, \ldots$ can have any integer value.
The second number $l = 0, 1, 2, \ldots$ up to the number $n - 1$.
The third number $m = 0, \pm 1, \pm 2, \ldots \pm l$.
The fourth or *spin* quantum number can have only two values $+1/2$ or $-1/2$.

These numbers, by the way, have no significance other than that stated above. By that I mean we do not do arithmetic with them but merely use them as labels to distinguish the different possible solutions of Schrödinger's equation. A different solution, involving a unique allowed energy, arises for each possible combinations of the first three quantum numbers."

"Let me be sure that I am following you." Beth interjected. "Suppose that we consider the solutions possible for the principal quantum number

$n = 3$. In this case, $l$ can have the values 0, 1, and 2 (= 3 − 1). Now, when $l = 0$, m can have the value zero, but when $l = 1$, m can have the values zero or plus-or-minus one. (See also Fig. 40.) Finally, when $l = 2$, m can equal to 0, ± 1, or ± 2. This gives us, let's see, one plus three plus five or a total of nine possible combinations. Right?"

"Right as rain, except that for each set of three quantum numbers that you enumerated, the fourth or spin quantum number can be plus-or-minus one-half."

"That's not difficult — so far." Beth began to sound a little less certain. "Let's go on to the Periodic Table of the elements."

"Very well." I agreed. "I can now give you the complete import of the Pauli exclusion principle: no two electrons in a single atom can have the same four quantum numbers. If you accept that as a golden rule, then you can construct the entire Periodic Table in a logical manner. The only other thing you need to know is that the atomic number $Z$ of an atom tells you how many electrons are needed to balance the charge of $+Ze$ on its nucleus. Care to try your luck?"

"Sure, but I better use a pencil and paper." Beth sounded confident once more. "We start with $n = 1$, for which $l$ and therefore, m must be zero. For $Z = 1$, the one electron in hydrogen than can have either spin number, plus or minus. For $Z = 2$, the two electrons in helium share the same first three numbers so that the spin of one must be plus and of the other minus one-half."

"As well done as these yummy waffles!"

"Please, let me go on." Beth was enjoying herself. "Having exhausted all possible combinations for $n = 1$, let's consider the possibilities when $n = 2$. Here, again, we have two possible combinations for which $l = m = 0$ and the spin is either positive or negative. So that lithium, with $Z = 3$, has two electrons for which $n = 1$, $l = m = 0$, while the third electron has $n = 2$, $l = m = 0$ and a spin number of, say, + 1/2. This uses up one of the two available spin numbers so that the fourth electron in beryllium (Be), with $Z = 4$, must use the second available spin number in combination with $n = 2$, and $l = m = 0$.

Next we have the case $n = 2$, $l = 1$. Now $m$ can be zero and plus-or-minus one so that we have a potential for six more electrons which should take us up to $Z = 10$ which is the atomic number of neon (Ne). How am I doing?"

"Very well!"

"So let me try one more row. When $n = 3$, $l$ can be 0, 1, or 2. We've already seen what happens when $l$ is 0 and 1, that gives us eight possible combinations which bring us up to $Z = 18$, the atomic number of argon (Ar).

When $l = 2$, we should get ten more possible combinations, yet the third row of the Periodic Table contains only eight atoms. Why is that?"

"Do you want a short answer or a longer one that will give you a clearer appreciation of this subject?"

"I've come this far, let's plunge ahead."

"Spectroscopists knew that atomic spectra, although unique for each atomic species, bore certain similarities." I was glad to explain. "So they classified their observed spectral lines into those that were 'sharp,' 'pronounced' (or 'principal'), 'diffuse', and 'further'. After quantum mechanics, it was possible to associate these lines with the quantum numbers of the electrons that gave rise to them so that the sharp lines, now designated by the letter $s$, were the ones for which $l = 0$. The principal or $p$ lines, correspond to $l = 1$. $d$, for diffuse, corresponds to $l = 2$, while $f$, for further, to $l = 3$."

"I think I see where this is headed." Beth was eager to have her turn. "The first two electrons in an atom must have $l = 0$ so they are both $s$ electrons. The next two electrons, for $n = 2$, similarly have $l = 0$ so they too are $s$ electrons. I'm guessing, therefore, that $s$ electrons produce 'sharp' spectral lines which is why the atom hydrogen (H) and lithium (Li), as well as helium (He), and berylium (Be), lie above each other in the same column of the Periodic Table."

"I told you that this would be fun!" I beamed. "You're quite correct in your analysis. In fact, the first two electrons are called 1$s$ electrons and the next two are called 2$s$ electrons in what is a very convenient short-hand notation."

"Let me go on." Beth was becoming really caught up in this. "The next six electrons will be designated 2$p$ electrons, and that's as far as we can go because we have used up all allowed $l$ values for $n = 2$. Is this what sets up the limit of atoms in a row?"

"Recalling that the column in the Periodic Table were devised to accommodate atoms having certain characteristics in common, what distinguishes the atoms at the end of each row is that they do not normally interact with other atoms." I explained. "In fact, they are known as the *inert gas* atoms because they do not easily combine with each other to form liquids nor do they enter into compound formation with other atoms. The underlying reason for this is that such atoms have used up all the allowed energy levels for a particular value of the principal quantum number $n$."

"I see." Said Beth. "That means that helium (He), having filled both allowed levels for $n = 1$, and neon (Ne), having filled the eight allowed levels for $n = 2$, are both inert gas atoms. What about the next row, however? We have ten additional possibilities for $1 = 2$, after accounting for the two $3s$ and six $3p$ electrons. Why then are there only eight atoms in the third row of the table instead of eighteen?"

"As the total number of electrons in an atom increases, their relative energies are increasingly affected by the other electrons present. "I tried to keep matters simple. "Such interactions cause the relative energies of the $4s$ electrons to be actually lowered as compared to those of the $3d$ electrons for which $l = 2$. Thus the nineteenth electron of potassium (K) can have a lower energy by becoming a $4s$ electron instead of trying to complete the sequence for $n = 3$ in a $3d$ state. And, of course, the second law of thermodynamics tells us that everything seeks to be in the lowest energy state possible."

"I've just discovered something else!" Beth was still excited. "All the elements in the first column of the Periodic Table have only one $s$ electron starting off the new row. Similarly, the elements in the second column have a pair of $s$ electrons, and those in the third column have one $p$ electron, and so forth."

"What you are noting is the basis for the organization that Mendeleyev deduced empirically." I observed. "We can now add one more step to our spectrographic notation and then the entire process of classifying the elements becomes nearly self evident. The number of electrons sharing the same $s$, $p$, or $d$ state can be indicated by adding a superscript to the letter symbol. Thus hydrogen's one electron is $1s^1$, while the electronic structure of helium is $1s^2$, of lithium, $1s^2 2s^1$, et cetera."

"Let me try do to a couple." Beth was really enjoying herself. "Let's see, aluminum (Al), with $Z = 13$, has the electronic structure

$1s^22s^22p^63s^23p^1$ and one more please, argon (Ar), with $Z = 18$, should be $1s^22s^22p^63s^23p^6$."

"Very good!"

"But I'm not clear as to what happens in the next row." Beth bemoaned. "From what you said, the energy of $4s$ electrons is lower than that of $3d$ electrons, so that potassium (K) and sodium (Na) have $4s^1$ and $4s^2$, respectively, following the same eighteen-electron assignments of argon. But what happens in scandium (Sc), $Z = 21$? Is the next electron a $4p$ or a $3d$ electron?"

"The energy of the $3d$ electrons is sufficiently lower than that of $4p$ electron so that the next ten elements proceed to fill the available ten $3d$ states." I explained. "That goes on until we reach nickel (Ni), $Z = 28$, whose electronic structure should be that of argon plus $4s^23d^8$, except that usually it isn't."

"This won't confuse me, will it?" Beth asked plaintively. "When I was doing so well?"

"I will not bore you with the details but what happens has very important consequences." I answered patiently. "As the number of $3d$ electrons in an atom increases, the energy differences between $3d$ and $4s$ electrons become smaller. This makes it possible for the outermost electrons in atoms of manganese (Mn) through nickel (Ni) to assume either state with equal ease. Thus the last ten electrons of an individual nickel atom are just as likely to be $4s^13d^9$. Inside solids, these outer electrons are affected by the electron clouds of neighboring atom as well, so their distribution among these two allowed states may be modified still further. In fact, we call manganese (Mn), iron (Fe), cobalt (Co), and nickel (Ni) *transition metals* for this reason.

The important consequence of all this lies in the ease with which the electrons can change their respective states. This accounts for the variety of properties that transition elements exhibit. For example, you may recall that these elements are ferromagnetic. Transition metals also can combine with other elements in a much larger number of ratios than most other metal atom can.

"What actually happens when one element meets another?" Beth was back pursuing her interest in the mating game of atoms.

## What Attracts Atoms to Each Other?

"To understand atomic bonding," I resumed the narrative, "we need to return to the Periodic Table (Fig. 42). The last element in each row has an electronic structure in which the available states for the last principal quantum number $n$ are filled. As we already noted, this is a very stable state for an atom so that the elements in the last column do not 'feel an urge' to interact with any other atoms. This is the case for neon (Ne) at the end of the second row. Going on to the next atom, sodium (Na), it has an 'inner' structure like that of neon plus one $3s$ electron outside this core."

"Is there a way for sodium to ditch that electron so that it too can be in the same stable state as neon?" Beth was quick to observe.

"When an atom loses an outer electron it is said to become *ionized*," I responded, "and the atom less that electron is called a positive *ion*."

"Is it positive because the charge of the atom's nucleus remain $+Ze$ while the number of negative electrons surrounding it has been decreased to $Z - 1$?" Beth speculated. "And does that mean that the next element, magnesium (Mg), can give up both of its $3s$ electrons and become a doubly charged ion?"

"Exactly." I continued. "We call such positive ions *cations* and their charge is referred to as their *valence*. Thus sodium has a valence of plus one and magnesium's is +2. In a similar fashion, elements toward the right end of each row in the Periodic Table can pick up an electron or two to attain the same outer electron structure of their nearest inert gas. When that happens, they become negatively charged *anions*. Thus a chlorine (Cl) ion has added one electron to have a valence of minus one, while sulfur (S) can pick up two electrons to have a valence −2."

"What happens when a positive sodium ion meets a negative chlorine ion?" Beth volunteered: "don't tell me — they form a sodium–chloride molecule."

"Not bad for someone who never studied physics." I kidded. "Having an inert-gas-like electronic structure, is like forming a sphere whose outside is uniformly charged negative or positive. As a result, each cation tends to attract to itself as many anions as can fit around it while each anion tries to surround itself with as many cations as it can. This leads to

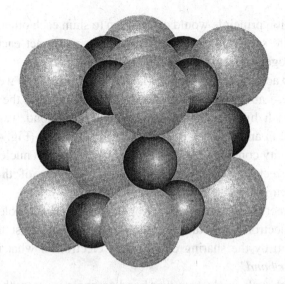

Fig. 43.  A part of the atomic array of chlorine (large) and sodium (small) ions in a sodium chloride crystal. Note that each ion is surrounded by six ions of the opposite kind so that their respective charges are equally neutralized. In actual salt crystallites, this array repeats itself about 10,000,000 times in all three directions!

the formation of a continuous solid (Fig. 43) rather than discrete smaller groups we call molecules."

"This must be the case because the electrostatic or Coulomb force exerted by a cation on an anion and *vice versa* is so very, very strong." Beth reasoned aloud. "We know that opposite charges attract each other. So, once a cation and an anion join together, any other charged ion coming close is instantly attracted to one or the other ion in the original pair. Because the electric charge is spread equally over each ion, this can go on and on until an array like the one in Fig. 43 is produced."

"That's essentially correct," I noted, "and we call this *ionic bonding.*"
"Too bad for the ions!" Beth was having fun again. "They can't have a monogamous relationship."

"Actually, some atoms manage to do just that." I observed. "Consider an atom of hydrogen with its 1s electron. What do you think happens when another hydrogen atom approaches it?"

"I suppose it depends on whether their respective spins are the same or not." Beth wondered aloud. "If the spins are alike, I would think that the

Pauli exclusion principle would cause them to shun each other. But if they have opposite spins, do the two hydrogen atoms attract each other and form a hydrogen molecule?"

"That's exactly what happens!" I kept marveling at how well Beth was grasping these abstract concepts. "Keeping in mind that the $1s$ electron surrounds each hydrogen nucleus in a spherical cloud, two hydrogen atoms joined in an $H_2$ molecule have a dumbbell shape (Fig. 44) with the electron density concentrated part way between the two nuclei."

"Is it correct to say that the electronic structure of the hydrogen molecule then is $1s^2$?"

"It is correct to say that each hydrogen atom in an $H_2$ molecule has an 'averaged' electronic structure $1s^2$." I replied. "This stable state is reached by each atom by the sharing of its single electron in what is called an *electron-pair bond*."

"How does such an electron-pair bond compare in strength to the ionic bond between oppositely charged ions?" Beth asked.

"Actually it is the strongest interatomic bond that we know." I replied. "Also called a *covalent* bond, its recognition is another triumph of quantum mechanics. You see, it turns out that electrons sharing the first three quantum numbers tend to pair with each other in an atom. Any unpaired electrons in one atom, therefore, are free to seek similarly unpaired partners in another atom with whom they can form electron-pair bonds."

(a)                                             (b)

Fig. 44.   Two hydrogen atoms can join to form an $H_2$ molecule by pairing their respective $1s$ electrons provided they have opposite spins. (a) The actual electron clouds surrounding the two hydrogen nuclei assume a dumbbell shape. (b) An alternate way of representing the electron-pair bond diagrammatically.

"That sounds like atoms other than hydrogen can form electron-pair bonds." Beth sounded uncertain. "But how do they go about it?"

"Let's return to the Periodic Table." I suggested. "Atoms of fluorine lack one electron from having an inert-gas configuration of neon. Suppose two fluorine atoms were to approach each other. By pairing the one outer electron that is unpaired within each atom with that of the other atom, they can form an electron-pair bond between two fluorine atoms in an $F_2$ molecule (Fig. 45). In this way, both fluorine atoms have an electronic structure like that of neon in a statistical sense."

"What happens when an atom has more than one unpaired electron?" Beth asked. "Can it form more than one covalent bond?"

"Absolutely." I replied. "Consider oxygen (O) with an electronic structure $1s^2 2s^2 2p^4$. If two of its $2p$ electrons should be unpaired, they can each form an electron-pair bond with unpaired electrons in another atom. Two oxygens can form two electron-pair bonds in an $O_2$ molecule. Or they can form one electron-pair bond with each of two different atoms. Such possibilities increase very quickly as we move toward the middle of a row in the Periodic Table."

"Is that how two hydrogen atoms pair up with one oxygen atom to form an $H_2O$ molecule?" Beth was excited by her discovery.

"Very good!" I had to smile at her enthusiasm. "Things gradually become more complicated as atoms having many electrons are considered. The nature of the interatomic bonds formed also can get somewhat blurred by a mixture of partly ionic with partly covalent forms of bonding."

"What about individual molecules? Do they have any way of bonding to each other to form a solid?" Beth's interest remained high.

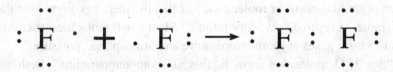

Fig. 45. Fluorine molecule formed by two fluorine atoms by pairing one of the $2p$ electron that is unpaired in an isolated atom. Note that this gives each atom a 'statistical' configuration $2s^2 2p^6$.

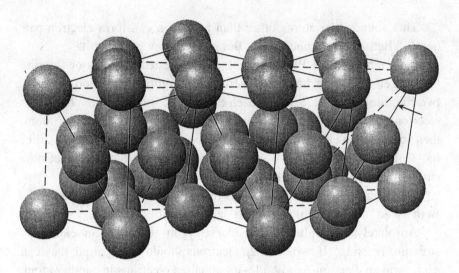

Fig. 46. A small portion of the array of carbon atoms forming a diamond crystal. Each atom is bonded to four other carbons by electron-pair bonds. The array extends for many tens of billions of atoms in each direction in any diamond large enough to be seen in a ring.

"First of all let me point out that covalently bonded atoms can form large crystals just like ionically bonded ions can." I went on with my explanation. "For example, carbon atoms join to form extensive networks in diamond crystals (Fig. 46). The extreme hardness and other properties of diamond all testify to the exceptional strength of the electron-pair bonds shared by neighboring carbon atoms. Small molecules, on the other hand, having attained inert-gas-like configurations in their constituent atoms by electron pairing, do not form solids easily."

"Like $H_2O$ molecules in water!" Beth really enjoyed her own perceptiveness, as I enjoyed watching her joy of discovery.

"Once all the electrons in an atom or a molecule are paired up, they shun other like atoms or molecules, just like the inert-gas atoms that they resemble." I resumed my story telling. "Most small molecules, therefore, tend to form gases at room temperature and atmospheric pressure."

"But $H_2O$ molecules form liquids at room temperature." Beth was quick to point out.

"That's right." I replied. "And the reason for this lies in the unique structure of a water molecule. Picture an oxygen atom as a kind of mushy sphere to which two hydrogen atoms are attached by electron-pair bonds shared between them. This localization of its electron leaves each hydrogen atom sort of exposed on its opposite side. This means that there are two partially unshielded positive hydrogen nuclei located at one end of each oxygen whereas, at the opposite end of the molecule, there is a slight excess of negative charge. The water molecule thus becomes an electric dipole with opposite ends that are positively and negatively charged."

"And electric dipoles can attract each other in the same way that magnetic dipoles can." Beth was right on the ball. "But why don't they form solids then?"

"Well, as you know, water does freeze at 32 degrees Fahrenheit or 0° centigrade." I responded. "Unlike magnets, whose north and south poles are permanently fixed in space, the locations of the hydrogen nuclei in a water molecule are less rigidly fixed at room temperature so that the bonding between neighboring $H_2O$ molecules is impermanent. This gives them more freedom to move about in the liquid state."

"So good old Democritus had the right idea after all." Beth was smiling. "Water *atomos* are slippery while iron atoms do have hooks on them. All we've learned from your quantum mechanics is that these hooks come in the form of electron-pair bonds."

"Oh how I wish that it were my quantum mechanics." I remarked wistfully. "But your general thesis is correct: our understanding of nature increases in a step wise fashion."

# Chapter Sixteen:
# Breakfast of O. J., Donuts,
# and Coffee

## *Surrounded by Fluids*

"I'm running a little late this morning." Beth announced. "So I'll be serving you just donuts and coffee. Would you also like some orange juice?"

"I'll squeeze the oranges." I replied. "I assume that we shall skip discussing physics this morning."

"Not necessarily. If you can keep it brief, I'd like to hear more about the slippery water molecules." Beth was ever curious.

"Well, as you know, most of the earth's surface is covered by water." I tried to be responsive. "Similarly, you and I are mostly made up of liquids. Moreover, the air that blankets us and the earth is also a fluid."

"Air is a fluid? I thought it was a gas."

"For present purposes, it is most convenient to divide all matter into solids and fluids." I explained. "A solid has a fixed size and retains its outward shape. It does this because the atoms in solids are bonded in ways that limit their ability to move past each other. Fluids, on the other hand, characteristically assume the shape of their container."

"But the atoms and molecules in liquids are still bonded to each other." Beth protested. "Otherwise wouldn't they fly apart the way the atoms in a gas do?"

"That's quite true. Atoms in liquids do form part-time bonds, as we've seen in the case of water. Liquids are also virtually incompressible, just like solids." I went on. "On the other hand, they do not retain their own shape but conform to that of their container. Their constituent atoms or molecules are also in perpetual motion so that they bombard the walls of the container just like gas atoms do. But you're also right that gas atoms do not bond with each other, even temporarily, so that they are totally free to move about at much larger speeds than their liquid counterparts."

"I never thought of it that way," Beth marveled out loud, "but if the atmosphere above us is a fluid, then are we living at the bottom of a sea?"

"Yes." I answered. "The air in roughly the first thirty-thousand feet above sea level belongs to the *troposphere*. All air above that belongs to the *stratosphere*. The density of the air gets lower as one goes higher so that about 99% of atmosphere is contained within a radial distance of about 120,000 feet. But, when a pilot announces that your flight is leveling off at forty-thousand feet, you are flying above 80% of all the air in the atmosphere below you."

"Whew!" Beth was grinning. "That's quite a load to carry when you get back on the ground. I mean all that air above us must be pressing down on us!"

"As a matter of fact it is and we call it one atmosphere of pressure." I proceeded to explain. "Humans normally aren't aware of it because all gases in our bodies, like the air in our lungs or in our blood, are at the same pressure. But when a diver descends in an ocean, the additional fluid or sea water above compresses the air inside the diver who is very conscious of the increased pressure."

"That's why divers have to come back up very slowy or else risk getting the bends." Beth spoke confidently. "Actually, when did people first recognize that they were so much like fish at the bottom of a sea of air?"

"Early on in the history of physics." I replied. "Shortly before he died, Galileo acquired a young assistant by the name of Evangelista Torricelli whom he directed to discover what made the water pumps work in the city of Florence."

"My favorite city in Italy!" Beth exclaimed. "Didn't they use a piston to create a vacuum above a water pipe in the well so that the water rushed in to fill the vacuum?"

"That's what everyone thought at that time." I went on. "But Torricelli reasoned that it wasn't the suction of the vacuum that was *pulling* the water up in the pump but the pressure of the atmosphere that was *pushing* the water to a height of 32 feet or 10.3 meters above the sea level."

"If he was a disciple of Galileo's," Beth wondered, "he must have had some scientific way of proving this conclusion."

"That he did. Torricelli knew that mercury was 13.6 times denser than water so that, if he was right, atmospheric air pressure should support a column of mercury that was 13.6 times less high than the 10.3-meter column of water."

"Oh I remember this from my general science class." Beth interrupted. "You take a long glass tube that is sealed at one end. Fill it with mercury. Put your finger over the open end and turn the tube upside down and insert it in open dish full of more mercury. Some of the mercury runs out but a column about 76-centimeter high remains above the dish (Fig. 47). Is that accurate?"

"You have an excellent memory." I beamed. "What Torricelli showed was that one atmosphere of pressure is equal to the pressure exerted by a column of mercury 76-cm high. Incidentally, he invented the very first *barometer* that proved invaluable in the later studies of gases by Boyle and others."

"Let's test my memory one step further." Beth was enjoying herself. "Wasn't it Archimedes who discovered that fluids exert a *buoyant* force on any body immersed in them that is exactly equal to the weight of the fluid displaced by that body? And did he actually jump out of a bath tub when he realized this and run naked through the streets shouting *eureka*?"

"I don't know about the reliability of that story, but people were much less prudish in those days so it might have actually happened." I replied.

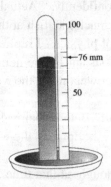

Fig. 47.   A simple barometer consists of a tube of mercury, sealed at its top, inverted in a dish of mercury. At sea level, the centimeter scale inserted in the mercury dish will indicate that the column is 76-cm high.

"Archimedes' principle, however, further illustrates why gases and liquids are both classified as fluids. Take an ordinary toy balloon and inflate it with air. When you let it go, the balloon will drop to the ground. If, instead, you inflate it to exactly the same size with helium, the balloon will rise rapidly high in the air. Can you tell me why?"

"A balloon of a given size displaces some amount of air." Beth sounded quite confident. "Since the balloon filled with air weighs more, because it contains compressed air plus the weight of the toy balloon itself, it is heavier and sinks to the ground. A balloon filled with the much lighter helium gas, weighs less than the same volume of air so the buoyant force of the air pushes the balloon upward."

"I couldn't have said it better myself!"

"As long as I'm showing off," Beth continued, "didn't Torricelli also discover that if you push on a fluid in a closed container, the pressure applied is distributed uniformly throughout that fluid so that a driver stepping on the brake pedal of a car transmits exactly the same pressure through the brake fluid to all four wheels of the automobile?"

"That was Blaise Pascal." I corrected her. "And Pascal's hydrostatic principle has very important consequences. One is the fact that liquids always seek their own level. Another is our ability to hoist a car weighing several tons using an imcompressible fluid and a modest-size air compressor (Fig. 48)."

"I believe it was the late Justice Oliver Wendell Holmes," Beth loved having last word, "who coined the phrase 'hydrostatic principle of human conduct' by which he meant to describe that people rose to their own level of competence in life."

"Or their own level of incompetence, if you believe in the latter-day Peter Principle." I quipped back. "But let me round out this morning's discussion by a few comments about solids."

"Before you go on, is a glass half full of a liquid or half empty?"

"That depends on whether the liquid is plain water or a first-rate wine."

## We Depend on Solids

"Well, the coffee depends on our cups to keep it from flowing all over the table," Beth remarked, "while the donuts retain their shape, including

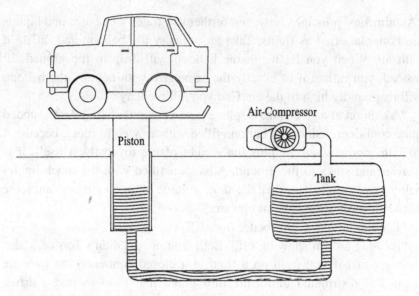

Fig. 48.   Pascal's principle enables operating a hydraulic lift at a corner gas station.

the hole in their middle, all by themselves. Yet both resist compression by an external air pressure, unlike the gas in a balloon. What causes the difference in their interatomic bonds?"

"Atoms in solids form permanent bonds with each other whereas those in liquids form temporary bonds that they keep breaking and reforming as the atoms move about."

"When you showed me the array of carbon atoms inside a diamond crystal (Fig. 46)," Beth asked, "was it based on the work of x-ray crystallographers like yourself?"

"Yes. By studying the way crystals scatter x rays directed at them, it is possible to determine what the internal array of atoms in solids is. Combined with the more recently developed quantum mechanics, this provided insights into the physical and mechanical properties previously limited to sheer speculation. It ultimately led to such discoveries as the *transistor* and truly revolutionized what became known as the electronic era that began with the discovery of the triode vacuum tube."

"Oh I remember vacuum tubes!" Beth exclaimed. "Whenever our early TV set stopped working, I took all the vacuum tubes out of its chassis and tested them in a tube tester at the corner drug store. By replacing the defective tube, I could actually make our TV work again! Nowadays even my smart professor-of-physics husband calls a repairman to service our TV. Is that the revolution that the transitor wrought?"

"That's the price we pay for what we call progress." I replied. "When I was a graduate student at M.I.T., there was a separate building filled with vacuum tubes that were the guts of a high-speed computer they called 'Whirlwind.' Today, your handheld calculator can perform most of the same mathematical operations at a much faster speed. That's the other side of progress."

"I guess you're right." Beth admitted reluctantly. "The computational ability of a transistor has enabled inordinate progress in communications and similar applications. Is the transistor you mentioned the same thing as the 'chip' I hear people talk about all the time?"

"Yes." I replied. "Transistors are made mostly of small chips of silicon whose atomic structure, by the way, is exactly the same as that of diamond (Fig. 45). Because a single silicon crystal or 'chip' contains about 100,000,000,000,000,000,000 silicon atoms, you can imagine that it should be possible to replicate the operations of a lot of vacuum tubes inside a single such chip. Although silicon for chip manufacture is plentiful and relatively inexpensive, increasingly more interest is being directed these days to other kinds of materials as well. In fact, a whole new field of materials science grew out of the need to combine the talents and skills of solid-state physicists, chemists, metallurgists, and related scientists, in a seemingly unending search for new materials and new ways to use known materials."

"Didn't you tell me that solid-state physics became the largest division of the American Physical Society?" Beth asked. "What about polymer science, is that part of physics or of materials science?"

"Polymers are, as you know, giant molecules composed of tens or even hundreds of thousands of atoms making up a single polymer molecule. The making of polymers, therefore, belongs to the realm of chemists." I explained. "The physics of polymers is, obviously, the province of polymer physicists while the characterization and utilization of polymers is what polymer scientists and engineers do. And in our interdisciplinary world, they all find a home in materials science."

"How did we wander from silicon chips to polymers?" Beth wondered out loud.

"Easily." I responded. "The discovery of the transistor stirred up the interest of huge numbers of physicists in the solid state. The U.S. reaction to the launching of a manned satellite by Soviet Russia caused our federal government to make a major investment in materials science research. Quick to capitalize on this opportunity, scientists proved that a concerted effort to understand the workings of nature pays off in the discovery of new ways to harness nature. The progress made in just the last thirty or so years has been truly phenomenal. Just think of some recent discoveries: the solid-state laser and what it and fiber optics can do in communications. The prospect of solid-state superconductors operating at or near room temperature ..."

"Hold up!" Beth weighed in. "I realize that I may have pushed the wrong button, but I only wanted to know why this growing emphasis on polymers?"

"The answer is miniaturization." I explained. "First to decrease the weight of components launched into space, but soon thereafter for all kinds of good reasons, the drive was on to cram ever more functions into ever smaller chips made from silicon. Polymer molecules, which, by the way, make up our nervous system and other body parts, contain many of the elements of communication networks already. What we need to figure out is how to tap into them under conditions that would allow us to control their operation from the outside."

"But polymers, which are commonly known as plastic, are not single crystals like silicon chips are they? I thought that transistors had to be made from single crystals."

"That's true," I explained, "but polymer molecules are so large that the negative and positive charges in the zillions of atoms comprising them produce electrostatic forces that tend to bind neighboring molecules fairly tightly to each other. Think of synthetic yarns like rayon or nylon, or plastic dishes. Even automobile bumpers are made from plastics. The molecules in these materials do not form highly periodic arrays found in crystals of silicon so that, speaking correctly, such solids should be called glasses. It turns out, however, that many of the properties attributed to crystals carry over to the less ordered arrays in glasses. This is why polymers and other glasses have been receiving increasing attention for possible applications in solid-state devices."

"I know that, if I ask you to explain what you just said, we'll be here until dinner time and I do have to run along. So I'll say thank you for today's lecture."

"And thank *you* for the delicious donuts and coffee!"

# Chapter Seventeen:
# Breakfast of Rice Krispies

## Who's Afraid of Radioactivity?

"Last night I thought up a dramatic opening for my chapter on nuclear physics. I wonder what you'll think of it?" I asked Beth one morning.

"On the morning of August 6, 1945, at 10:45 Eastern Daylight Saving Time, the White House announced that 'Sixteen hours earlier, an American plane dropped a single atomic bomb on Hiroshima, an important center ....' In these words, a war-weary world learned of President Truman's fateful decision to hasten the end of World War II."

"It will certainly catch the reader's attention." Was Beth's response. "Pardon the irony, but it seems to fit in with a breakfast that goes snap, crackle, and pop!"

"True, it's not a joking matter." I sensed that Beth wanted to change the subject. "Whatever we now think of Harry Truman's decision made half a century ago, the origins of the bomb actually can be traced to a misguided experiment conducted one whole century ago. Antoine Becquerel, the son and grandson of physical scientists, had wrapped a phosphorescent mineral containing uranium in black paper. He stored it next to a photographic plate in the drawer of his desk while waiting for the sun to shine on his Paris window sill."

"Was Becquerel trying to establish the role of sunlight in causing the phosphorescence of this mineral?" Beth asked.

"That's right, he was." I replied. "He already had tried wrapping a single thickness of paper around the mineral and now was waiting to see what effect several thicknesses of black paper would have. To his surprise, the photographic plate that had been in his drawer for several days, but in the sunlight on his window sill for only one day, developed a spot that was much blacker than that produced in previous exposures to the same amount of sunlight. Since the sun did not cause it, he was forced to conclude that something else was responsible. Could it be the phosphor?

Heating, freezing, grinding, or dissolving the mineral would not eliminate the source of the radiation. Becquerel finally came to the conclusion that it was the uranium present in the mineral that gave rise to the radiation causing the blackening of the photographic plate."

"I remember another event relevant to your story that also took place in Paris at that time." Beth chimed in. "A Polish student, Maria Sklodowska, married her French physics professor, Pierre Curie. When they learned of Becquerel's discovery, they set out to pursue their own researches in what Mme. Curie subsequently named *radioactivity*."

"You're quite right. The Curies quickly connected the radioactive emission to the presence of uranium and not to the phosphorescence of a mineral." I went on. "In 1903, they shared the third Nobel Prize in physics with Becquerel for their respective discoveries in radioactivity."

"Mme. Curie was the really famous member of this husband-and-wife team." Beth noted with pride.

"Purely a matter of chance." I parried. "When he married Marie, Pierre Curie was already well known among physicists for his researches on the magnetism of materials. In fact, we call the transition point at which a substance becomes ferromagnetic the *Curie temperature*, in his honor. If he hadn't been killed by a reckless carriage driver while crossing a street in Paris, who knows what contributions he might have made. As it was, Mme. Curie continued their researches of radioactivity, discovering radium and many other things that led to her winning a second Nobel Prize in 1911, this one in chemistry. You can take pride, moreover, in the fact that, not only was she the first woman to win a Nobel, but the first scientist of either gender to win two!"

"Did Mme. Curie contract any illnesses from her prolonged exposure to radium?" Beth asked sympathetically.

"As you can imagine, the early investigators had no idea of the dangers to their persons that this newly discovered radioactivity might entail." I responded. "It is now believed that Mme. Curie died of leukemia at age sixty-seven, but the medical profession at that time had not yet developed any diagnoses of radiation sickness. We do know that her laboratory notebooks preserved her radioactive fingerprints for many years after her death."

"What was the nature of the radiation that the radioactive elements emitted?" Beth wanted to know.

"Following in the footsteps of Röntgen," I replied, "Becquerel first tested how penetrating the emitted radiation was. He discovered that there were two kinds present. One was easily absorbed by a few layers of paper or about an inch of air while the other was much more penetrating and was not totally absorbed even by several thicknesses of metal foil. These, by the way, became subsequently known as *alpha* and *beta* rays, respectively."

"Wasn't the name-giver for these rays Rutherford? And hadn't he moved to Montreal in Canada where he took up the study of the other radioactive element, thorium?" Beth contributed.

"Quite so." I agreed. "It was in Canada that Rutherford discovered what we now call the *half-life* of a radioactive substance."

"I know what that is!" Beth exclaimed. "The radioactivity declines to one half of its previous value during some time interval. An equal time later it drops to half of that value. Wait the same length of time once more, and the remainder will have lost half of its radioactivity, and so on, until what's left is too small to measure. But what were the alpha and beta particles like and how did they differ?"

"Becquerel observed that the more penetrating beta rays could be deflected by magnetic or electric fields not unlike the observations on cathode rays made by Thomson earlier." I explained. "This was confirmed by Rutherford who also succeeded in deviating the alpha rays in a very strong magnetic field. He established further that the electric charge of alpha rays was about twice that of hydrogen. Combining this with geological evidence that helium gas seemed to accompany uranium- and thorium-containing minerals led Rutherford to assume that alpha rays were actually positive helium ions. More convincing proof of this came eight years later.

Finally, Rutherford also discovered a third kind of radiation that could not be deviated by any magnetic or electric field. These extremely penetrating rays could pass through more than an inch of lead! Following the Greek alphabet, he named them *gamma* rays."

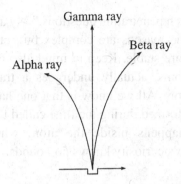

Fig. 49. Three kinds of particles emitted by a radioactive source: alpha particles are helium atoms stripped of their two electrons. Beta particles are high-speed electrons. Gamma rays are very high-energy photons.

"Can you summarize for me how these three kinds of emissions compare?" Beth asked for clarification. "Are any of them part of the electromagnetic spectrum like x rays?"

"If you can picture a magnetic field at right angles to this sheet of paper," I suggested, "and a radioactive source at the bottom (Fig. 49), then the fast-moving alpha particles, now known to be positive nuclei of helium atoms, will bend off to the left. The beta particles, being high-speed electrons, bend to the right, and the gamma rays, being high-energy photons, do not bend at all."

"In that case, only the gamma rays are part of the electromagnetic spectrum." Beth answered her earlier question herself. "But what happens to the element after it emits such particles?"

"This remained an elusive question until Rutherford, assisted by Frederick Soddy, developed his *disintegration hypothesis* in 1903." I replied. "Based on their observation of the decline in the radioactivity of thorium, they established that radioactivity involved not only the emission of one or more kinds of radiation but also a simultaneous transformation of the 'parent' atom into a different 'daughter' atom. If it too was radioactive, the daughter atom could undergo a transformation into a third kind of atom by the emission of radiation as well."

"And each radioactive atomic species has its own characteristic half life." Beth observed. "What I don't understand is why these half lives can range so widely."

"You tend to ask such penetrating questions." My attempted pun passed unacknowledged. "The reasons are complex but, clearly, atoms having long half lives are more stable. Keep in mind also that which particular one, in a pile of atoms, actually undergoes a transformation at any moment is quite random. All we know is that one half of the radioactive atoms will have transformed during the time called their half life."

"What actually happens inside the atom when it undergoes a transformation?" Beth's curiosity knows no bounds.

## What's Inside the Nucleus?

"As our understanding of radioactive processes increased," I resumed my story, "it became apparent that the transformations were actually taking place within the nucleus of an atom. This enabled Soddy to show that two atoms of the *same* element could have the same atomic number or electric charge on the nucleus while having two different nuclear masses. In fact, he extended this observation to include elements that were not radioactive as well. He named them *isotopes* of the same element."

"Is that what won Soddy a Nobel Prize in chemistry?" Beth asked but then answered her own question. "I seem to recall this tidbit from reading that the several different kinds of isotopes can make up a single element found in nature. Before then I always thought of the ninety-two elements in the Periodic Table as being unique."

"At first, the additional elements that are produced by radioactive decay did cause some skeptics to question the Periodic Table," I went on, "but Soddy showed that isotopes were quite common in nature and did not violate the Mendeleyev classification scheme. Even an inert gas like neon consists of atoms having different masses. From the measured mass of neon atoms we can tell that 91% are the isotope neon-20 while 9% are the heavier isotope, neon-22."

"I've seen that designation before," Beth stated, "like uranium-238, for example. What is the significance of the number following the element's name?"

"We express atomic masses in what we call the atomic mass unit or *amu*," I replied. "The number appended to the element is the number of these amu that the nucleus weighs. A nucleus of U-238 weighs 238 amu while that of U-234 weighs four amu less."

"What accounts for these differences in the weights of two nuclei?" Beth asked.

"The fact that nuclear weights tend to be very nearly whole numbers of amu, combined with the fact that their transformations produce nuclei differing by integer multiples of amu, led Rutherford to speculate that a nucleus is made up of what we now call *nucleons*, each weighing one amu." I went on: "consider the radioactive series radium-226 transforming to radon-222, which, in turn, transforms to polonium-218, which becomes the nonradioactive or stable lead-214. An alpha particle, the nucleus of helium-4, is emitted by each radioactive element in this series."

"This may explain why the weight of each atom in this series decreased by 4 amu," Beth observed, "but what is the makeup of a nucleus to start with?"

"Starting with hydrogen," I explained, "we have a nucleus whose weight is very nearly 1 amu and electric charge is the same as that of one electron, except that the charge is positive. We call this nucleon a *proton*.

Helium, the next element, has a positive charge of plus two electrons which is needed to bind its two negative electrons to the nucleus. It seemed reasonable to posit, therefore, that a helium nucleus contains two protons."

"But its nuclear weight is four!" Beth interjected. "Similarly, I remember that oxygen has an atomic number of eight and an atomic weight of sixteen. If its nucleus contains eight protons to provide the positive charge of +8 necessary to bind its eight electrons, what accounts for the other 8 amu in its nuclear weight?"

"At first Rutherford reasoned that the oxygen nucleus contained sixteen protons, to account for its weight, and eight electrons, to account for its net charge." I responded. "But Niels Bohr pointed out to him that confining eight electrons within the dimensions of a nucleus would require inordinately high energies. They agreed, therefore, that remaining eight nucleons in oxygen had to be chargeless. They also decided to name them tentatively — *neutrons*."

"How long did it take physicists to isolate a neutron? And who did it first?" Beth easily sensed what was likely to follow.

"As you can imagine, an active program to find a neutron was started in Rutherford's laboratory." I replied. "But it was not the only place. Although they did not appreciate it at the time because of the crudeness of the available detection equipment, the daughter of Mme. Curie, Irene, together with her husband Frederic Joliot, had detected neutrons but mistakenly thought that they were extremely penetrating gamma rays. James Chadwick, a former student of Rutherford, correctly reinterpreted their observations in terms of the emission of the elusive neutrons. This quickly proved to be a boon to the interpretation of many other radioactive transformations."

"And when did this occur?" Beth, if anything, liked to keep her chronology straight.

"Rutherford actually demonstrated the ejection of protons from a nitrogen gas bombarded by alpha particles in 1919." I replied. "His conversations with Bohr took place after that. It was not until 1932 that Chadwick's Nobel Prize-winning paper came out."

"From what you told me at a previous breakfast," Beth was thinking hard, "if an atom were the size of a football field, its nucleus would fit inside a single pea. How is it possible then for eight protons and eight neutrons to fit inside a single oxygen nucleus? I would think that the electrostatic repulsion between the positive protons would cause them to fly apart, not stick to each other!"

"Well reasoned!" I glowed with pride. "The electromagnetic force was, by far, the strongest of the only two forces known to exist between objects on earth. This left the physicists little choice but to postulate the existence of a *strong nuclear* force that was capable of overcoming the Coulomb repulsion of neighboring protons while binding them together with the neutrons inside the tight little nucleus."

"If there exists a strong nuclear force," Beth was quick to ask, "is there a weak nuclear force also?"

"Indeed there is." I replied. "The weak nuclear force plays a role in the decay of a nucleus during a nuclear transformation. In anticipation of your next question, let me tell you that the weak nuclear force is nearly a trillion times weaker than the electromagnetic force. That's still 100,000,000,000,000,000,000,000 times stronger than the gravitational force of attraction. The strong nuclear force is about 100 times stronger than the electromagnetic force but it acts over extremely short distances only. By the time you look outside the tiny nucleus, the strong nuclear force has dropped off to virtually zero strength."

"You actually anticipated two of my questions." Beth sounded somewhat disappointed. "The reason why we are not aware of the nuclear forces is that they do not exist outside a nucleus. By the way, is this what gives a nucleus the energy that we make use of in nuclear power plants?"

"When a heavy nucleus like that of uranium-238 breaks up into two smaller nuclei weighing closer to 100 amu each, then some of the binding energy per nucleon is liberated and can be used to do useful work."

"Or to blow up unwary city residents." Beth wryly seized the opportunity to express her opposition to warfare. "So tell me how nuclear energy is released."

## *For Better or For Worse — Nuclear Energy*

"When a uranium-238 nucleus emits an alpha particle," I began my explanation, "it becomes a thorium-234 nucleus."

"What about the positive charge carried away by this nucleus?" Beth asked. "Doesn't the helium emitted also remove two protons thus reducing the nuclear charge of what's left by two?"

"Well done!" I glowed. "The nuclear charge, which equals the atomic number, decreases from 92 in uranium to 90 in thorium. It turns out that one half of the newly formed thorium-234 nuclei emit an electron within 24.1 days and transmute to palladium-234 whose atomic number is 91."

"I see." Said Beth. "The emission of a negative electron does not change significantly the mass of the nucleus but makes it more positive by +1 so that the new nucleus has a charge of +91."

"What's more," I added, "within about one minute, half of the palladium-234 isotopes emit another electron, changing the nuclear charge to +92. This, of course, is the atomic number of uranium so that we get another isotope of uranium, U-234. This isotope is quite stable, by the way, with a half-life of a quarter of a million years."

"Didn't you just tell me that Bohr convinced Rutherford that a nucleus couldn't hold an electron?" Beth protested. "Where then do these beta particles come from?"

"That's a very good question that puzzled physicists for some time." I replied. "As usually happens when presented with certain data that defy any other possible explanation, it is necessary to invent a new one. In this case, it was assumed that the electron was created simultaneously with the conversion of a neutron to a proton within the nucleus. Thus one of the neutrons in thorium-234 becomes a proton, increasing the charge to +91 from +90, while the negative electron created to conserve charge neutrality is ejected from the nucleus in the form of a beta ray."

"So this is sort of like the creation of an electron–positron pair by a high-energy gamma ray in the earth's stratosphere." Beth ruminated. "What about the mass–energy conservation in all these processes?"

"You are now getting close to answering your earlier question of how nuclear energy is released." I went on: "when one does a careful calculation of the mass–energy balances involved in the release of an alpha particle, the masses of the parent and daughter atom differ by exactly the mass of a helium nucleus plus the kinetic energy of the emitted alpha. A similar calculation in the case of beta emission also appears to work except for one puzzling thing. The beta particles that are emitted by the same kinds of atoms, as first observed by Becquerel, do not carry away exactly the same amount of energy!"

"If the energy carried off is different, aren't you violating the basic principle of energy conservation?" Beth sounded concerned.

"Indeed!" I was quick to agree. "This dilemma led Niels Bohr to suggest that, possibly, in nuclear reactions energy is conserved only in a statistical sense. In this case, the *average* energy carried away would still satisfy the overall energy conservation requirement.

Although this suggestion seemed to agree with some very precise energy measurements of the emitted beta rays, it bothered the sense of physical order in Wolfgang Pauli's mind. So he suggested that some mysterious other particle carried off any 'missing' energy."

"Are you saying that when all the masses and energies of the 'known' particles are added up, any unaccounted for energy can be assigned to an unknown particle?" Beth sounded bewildered. "Isn't that a bit hokey? Even for a theoretical physicist?"

"Not when he is convinced that the laws of energy conservation brook no violations whatever." I stated firmly. "There is an interesting aftermath to this story, by the way.

Because his postulate preceded Chadwick's discovery, Pauli named the mysterious particle a neutron since it had no electric charge. When Enrico Fermi, a truly exceptional blend of theoretical and experimental physicist, was lecturing a couple of years later in Rome about Chadwick's discovery of the neutron, a student asked him whether it was the same as Pauli's neutron. Fermi replied: oh no, Pauli's was much smaller. In Italian, the diminutive of neutron came out *neutrino* and this became the name of the mysterious particle."

"You physicists are really something!" Beth sounded exasperated. "You conduct an experiment. Then you can't account for what you have observed, so you blithely invent an imaginary particle and all is at peace in your world."

"Your protest is justified." I agreed. "But as a behavioral psychologist, you know very well the power of reinforcement. Previous bold and imaginative postulates deemed essential, although unprovable at the time they were made, later turned out to be correct. Is it really so surprising, therefore, that subsequently physicists would feel challenged to make similar attempts?"

"You're trying to tell me in your gentle way that neutrinos have actually been observed."

"As a matter of fact, it took a while, but by 1955 some very large absorbers and detectors especially arrayed for this purpose did manage to record the passage of the very elusive neutrinos."

"Didn't some scientists at the Los Alamos National Laboratory recently claim to have detected neutrinos that had some mass?" Beth asked.

"Yes they did and this may have some important cosmological implications." I agreed.

"But well before a neutrino had been first detected in 1955," I resumed, "and while still in Rome, Fermi worked out the complete theory of beta decay by involving Pauli's neutrino. This earned him the 1938 Nobel Prize. Fermi also carried out many experiments in which he bombarded heavy elements with neutrons in order to induce nuclear transmutations in them. In the course of these experiments he discovered that lower energy neutrons were actually more effective for this purpose than high-energy ones. This discovery enabled him years later to mastermind the first sustained nuclear chain reaction under the football stadium in Chicago."

"Wait a minute," Beth interjected. "Where did the idea of bombarding a nucleus to induce a transmutation come from? I thought that the transmutations occurred naturally as a result of spontaneous emission of particles by a radioactive element."

"You now have come to the birth of the atom bomb." I replied. "It took place, fittingly enough, in the experiments that the Joliot-Curies conducted but did not fully understand."

"I'm torn between my innate curiosity and my doubts about the atom bomb." Beth wavered. "What was it that Mme. Curie's daughter had missed?"

"The experiment that Chadwick analyzed involved a beryllium-9 nucleus absorbing an alpha particle." I resumed. "He correctly showed that the result was a carbon-12 nucleus plus a gamma and a neutron given off in the collision. Somewhat later, this suggested to Fermi that neutrons also might serve as bullets to induce transmutations in nuclei. His successful pursuit of such transmutations ultimately led him to recognize that a low-energy or 'slow' neutron was more easily captured by a nucleus of a stable atom. This turned the nucleus into an unstable isotope that, typically, emitted an electron to become the more stable nucleus of the next heavier element in the Periodic Table."

"And when the electron was emitted, the absorbed neutron was converted into a proton." Beth announced proudly.

"That's right." I added: "the careful analysis of his results enabled Fermi to postulate the existence of the 'weak' nuclear force to account for the process by which the electron was emitted. This process, by the way, is called *beta decay* and takes place in most elements bombarded by neutrons."

"Let's get back to the birth of the bomb." Beth insisted. "I really don't know why we should, except that we started on this trail when you first sat down for breakfast."

"You may be interested to learn that a woman played the role of a catalyst by pointing the way."

"Did she win a Nobel Prize for that?" Beth asked skeptically.

"As a matter of fact, she didn't," I replied, "although many believed she deserved it."

"If not the one for physics, what about the one for peace?" Beth added sarcastically.

"Lise Meitner was collaborating with her laboratory's director, Otto Hahn, in replicating some of Fermi's transmutation experiments when Adolf Hitler's rise in power made her decide to abandon Nazi Germany for a more hospitable Sweden. Hahn, who valued highly her collaboration, kept Meitner informed of his progress. It was she who first realized that the breakup of a uranium atom into isotopes of krypton and barium should be accompanied by the release of energy to account for the mass differences between the starting and final products."

"There may be a shortage of women in science, particularly in physics," Beth observed, "but the ones who persevere have a way of distinguishing themselves."

"Lise Meitner communicated her discovery to Niels Bohr through her nephew, Otto Frisch, who also was a physicist and confirmed her analysis." I resumed my story. "Bohr received the information on his way to a conference of theoretical physicists in Washington, D.C. Some of the details of what took place at this conference differ according to the memories of those who were present, but this much is clear: the expatriate Russian physicist, George Gamow, chaired a session in which he invited Bohr to describe the German experiments and Meitner's interpretation of them. The original session topic was abandoned as the excited physicists proceeded to analyze the significance of the breakup of a uranium nucleus into two nuclei plus the release of neutrons and additional energy."

"Did the conferees realize the full implications of this discovery?" Beth wondered out loud.

"They realized that the energy released could conceivably be harnessed for good or for evil purposes." I explained. "But they also agreed that a more effective source than U-238 would be needed. Thus one of the important results of this meeting was the conclusion that the breakup of a uranium nucleus was much more likely in the isotope U-235 than in the more stable U-238. They also realized that U-235 made up less than 0.7% of the uranium present in any uranium compound. The separation of this isotope from the other 99.3% uranium atoms appeared to be an insurmountable obstacle."

"Oh that it only remained inseparable!" Beth intoned cheerlessly.

"The impetus for inventing a way to increase the amount of U-235, or to 'enrich' natural uranium, actually first came up at this epochal meeting in January of 1939. The physicists present realized that it should be possible to establish a chain reaction with this uranium isotope, whereas it wouldn't work with just U-238. It is noteworthy that Bohr later opined that such a project would be utterly impractical 'unless you turn the whole United States into one huge factory.'"

"What, exactly, is a chain reaction?"

"When U-235 absorbs a neutron, it transmutes into U-236. This, in turn, almost instantly splits into the nuclei of barium-141 and krypton-92. The mass difference between the uranium atom and the resulting two atoms of barium and krypton consists of two or three neutrons emitted plus the energy liberated in this reaction."

"And these neutrons are absorbed by other U-235 atoms present which split up and liberate two or three neutrons each, which can be absorbed by still more U-235 nuclei, and so on and on." Beth had figured it out. "But, if the presence of U-235 is sufficient to start a chain reaction, why doesn't any pile of uranium do the trick?"

"Remember that less than one % of uranium consists of the isotope U-235." I explained. "Moreover, U-238 also tends to absorb neutrons. This is why the separation of U-235 from U-238 was such a critical issue.

By the way, the idea of sustaining a chain reaction had first occurred to one of the several Hungarian expatriate physicists to figure prominently in the development of nuclear energy in the United States. Leo Szilard actually got into a fierce debate with Lord Rutherford who refused to accept the idea that a nuclear chain reaction could be sustained or energy liberated thereby. Upon returning to the USA, Szilard's reaction was to take out a patent on the operation of a chain reaction. After the end of World War II, the U.S. government paid him $20,000 for the rights to that patent."

"What actually happens in a uranium nucleus when it captures an extra neutron?" Beth wanted to know.

"According to a model first proposed by Gamow and later refined by Bohr," I responded, "the nucleons inside a heavy nucleus are crowded together into an extremely dense ball. Those inside this ball are surrounded by other nucleons on all sides so they feel the attractive strong force of their neighbors equally from all directions (Fig. 50). Those lying

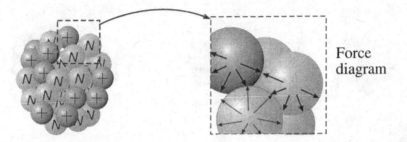

Force
diagram

Fig. 50. An atomic nucleus is an extremely dense packing of protons and neutrons held together by the nuclear strong force. The outermost nucleons feel radial and tangential forces not unlike those responsible for the surface tension in liquid drops. This surface tension is proportional to the square of the radius while the size of the drop increases in proportion to the radius cubed. If the radius exceeds some critical value, the surface tension can no longer maintain the drop as a single entity.

along the outer surface are pulled inward and sideways only, in the same way that water molecules in a water drop are. This is called *surface tension* and it is responsible for the cohesion of a water drop as well as a nucleus.

When an additional water molecule enters a water drop, or an external nucleon penetrates a nucleus, it loses its excess kinetic energy by successive collisions with the other particles present. The overall energy gained by the nucleus may cause it to emit a single particle, as in the case of radioactive decay, or it may lead to a collective excitation of all nucleons present. Such an excited nucleus then can undergo *fission* by splitting into two fragments whose binding energy per nucleon is lower so that they are considerably more stable (Fig. 51)."

"And any mass difference is converted into the energy which is carried off by the fission products." Beth chimed in. "What exactly is meant by the 'critical size' that I remember hearing about in connection with a sustained nuclear reaction?"

"Upon fission, each U-235 nucleus emits, on average, 2.5 neutrons which can travel about 3.5 inches before being absorbed by some other U-235 nucleus. Clearly, if the pile of uranium atoms is much smaller than that, most of the neutrons produced will escape from the pile without causing any new fissions. Increasing the size of the nuclear pile increases

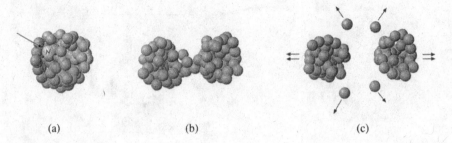

(a)                    (b)                    (c)

Fig. 51.   After a neutron has penetrated into a heavy nucleus (a), the agitation of the other energized nucleons may be reduced by the division of the excited nucleus into two nearly equal halves (b). Having a lower binding energy per nucleon, the two halves fly apart with an attendant release of any 'excess' neutrons and energy (c).

the likelihood of neutron capture so that the minimum size needed to sustain the chain reaction has been named the critical size."

"What happens if the size of the nuclear pile becomes much larger than this critical size?" Beth asked next.

"In that case," I explained, "the reaction rate increases progressively as each fissioned nucleus releases either two or three neutrons. To control the chain reaction, therefore, it is necessary to control the size of the regions within which the chain reaction can proceed. As late as the 1950's, the size and configurations of these regions was a closely guarded secret in a nuclear pile I visited at the Oak Ridge National Laboratory."

"How were these regions controlled?" Was Beth's logical next query.

"The elements boron and cadmium are very efficient absorbers of neutrons." I explained. "Rods of cadmium are inserted into a pile at intervals that determine the size of the active regions. Called *control rods*, they can be inserted or withdrawn to control the operation of a nuclear reactor by limiting the rate at which fission proceeds."

"How did the builder's of the first pile realize all of these ramifications?" Beth sounded somewhat awed.

"On December 2, 1942, Fermi, Szilard, and their many collaborators watched the very first nuclear reactor begin its operation in the University of Chicago's football stadium. As the cadmium control rods were withdrawn, the pile went critical and started generating heat. Just in case something were to go wrong, two volunteers stood atop the pile under the bleachers of Stagg Field holding buckets filled with cadmium solution, ready to douse any nuclear fire below them. After successfully generating heat for two hours, the reactor was shut down by the reinsertion of the cadmium rods."

"This whole episode reminds me of the tale about the sorcerer's apprentice." Beth observed excitedly. "The risks that those physicists took! What if the chain reaction had not turned out to be controllable? I shudder at the thought!"

"Fortunately, the first sustained chain reaction proceeded without any mishap. The importance of this event can be gauged by the TOP SECRET message that was telephoned to Washington. Referring to Enrico Fermi, who'd masterminded most of this operation, it simply said: 'The Italian navigator has set foot in the new world.'"

"An apt choice, since what followed certainly can be likened to the discovery of America." Beth chimed in.

"The purpose of the chain reaction carried out in Chicago was not limited to verifying the predictions of the theoretical physicists." I observed. "Another important result was the production of plutonium-239. Created by the transmutation of the highly abundant U-238, after it captures a neutron, plutonium also undergoes fission with the release of multiple neutrons. Because it is chemically different from uranium, plutonium can be separated from its parent U-238 far more easily than can the isotope U-235. For this reason, it was plutonium that was actually used in constructing the first atomic bomb."

"Do you know how to make an atom bomb?" Beth asked in an awed voice.

"To make a nuclear explosive," I responded, " it is necessary to merge two subcritical masses with each other so rapidly that the mass produced becomes supercritical and the fissioning nuclei release their energetic fission products virtually all at the same time. I know that this can be done by packing ordinary explosives around the fissionable materials but the details are a not-so-well-guarded secret of federal agencies."

"What is the nuclear fuel that is used to generate electricity?" Beth inquired.

"Although the use of plutonium would have many advantages, including the ability to generate more fuel than was being consumed in generating heat," I replied, "plutonium fuel presents an engineering challenge that the United States has been unable or unwilling to meet. One very serious drawback is that plutonium is highly toxic. A relatively ill informed public, including our elected representatives in Washington, moreover, have brought further developments in this area to a virtual standstill."

"You don't believe that this is a good thing?" Beth asked with concern. "What about the breakdown in Three Mile Island and the far more serious disaster in Chernobyl in the Ukraine?"

"I don't minimize the risks inherent in any process involving the generation of huge quantities of energy." I explained as calmly as I could manage. "The way to minimize risks, however, is to learn more about their underlying causes. We have had many malfunctions of nuclear reactors in the United States during their sixty-year history, without the loss of a single life or serious radiation exposure of populations nearby a reactor. Moreover, such accidents and the construction and testing of atom bombs since 1951 has released increasing amounts of carbon-14 into the atmosphere. As I will describe at a future meal, it is possible to utilize the known increase in dating wines, human tissues, and in other interesting applications.

"Well, you are right that there are many needless risks that we face every day." Beth agreed. "I don't find that a compelling argument, however, in defense of any one of them. What about nuclear fusion as opposed to fission. How does it work and is it safer?"

"Let me separate your two questions and answer them one at a time." I replied. "By splitting a heavy nucleus like uranium into two lighter ones, we gained energy from the lowered binding energies in the latter and their mass differences. By fusing two very light nuclei into a heavier one, we can also gain energy provided that the mass of the heavier nucleus is less than the sum of the two lighter nuclear masses. This turns out to be the case for several different possible atomic combinations. In our sun, for example, hydrogen nuclei, fuse to produce helium whose mass is less than that of the four protons fused in the reactions. Such a process goes on continuously and accounts for the sun's extremely high temperature of about 27 million degrees Fahrenheit. Other fusion processes can involve carbon atoms and are believed to account for the even higher temperatures reached by more distant stars."

"Well, if we know how the stars do it, why don't we duplicate the fusion process on earth?"

"That's exactly what several groups have been trying to do." I responded. "The difficulties in this process are that the hydrogen nuclei have to be heated to extremely high temperatures to give them sufficient energy to fuse and they have to be forced into very close proximity to each other. Achieving this high density in what is sometimes called the fourth state of matter, or *plasma*, and raising its temperature sufficiently is what it's all about. Capricious financing decisions by Congress have not helped speed up the development of a practical fusion process. To date, the energy consumed in producing fusion has exceeded that released by the fusion process."

"Isn't that what the laws of thermodynamics would predict for any attempt to get something for nothing?"

"But we are not trying to get something for nothing. The laws of energy conservation are not violated by the conversion of mass to energy. As I've said, this fusion process goes on in the sun and is responsible for our ability to live in the sunshine on earth."

"Is it the risks inherent in nuclear fusion that have concerned our Washington representatives?" Beth innocently inquired.

"I doubt it. Look at the way they keep appropriating money for a space station, year after year, even though it has been shown repeatedly to have very limited value beyond providing jobs and profits for its builders." I replied. "On the other hand, even if we succeed in generating more energy by nuclear fusion than we consume, there will remain technical problems to solve before it can be commercialized."

"That sounds like a preamble to the usual claim of scientists for ever more funding to support their research." Beth observed. "No wonder that Congress has lacked enthusiasm for this enterprise."

"It may have its own problems and even dangers," I interjected quickly. "Unlike fusion reactors, however, fission processes on earth are not self sustaining so that there is no risk comparable to a nuclear melt down by an out-of-control chain reaction."

"What about radioactive fall out?"

## *Our Nuclear Legacy*

"On earth, humans are exposed daily to electromagnetic radiation in the form of infrared, ultraviolet, and x radiation from the sun, as well as to radiation brought in by cosmic rays from more distant sources and that emitted by radioactive minerals in the earth's crust." I explained. "All artificial sources, including medical and dental x rays, nuclear power plants and military testing of nuclear weapons probably have doubled that amount for an average U.S. citizen (Fig. 52). Although the effect may be cumulative, it is not clear what special risks this radiation exposure presents to humans."

"How do we really know what is and what isn't an acceptable radiation level for humans?" Beth wanted to know.

"There is no unique answer because it intimately relates to how any particular radiation is absorbed by living tissue." I replied. "Hiroshima and Nagasaki in Japan have provided inadvertently a human laboratory for measuring the after effects of unusually high exposures to nuclear radiation. Combined with other ongoing studies around the world, we continue learning about the effect of different kinds of radiation on humans and how to measure and prevent harmful exposures."

"Has anything good come out of any of this?" Beth asked peevishly.

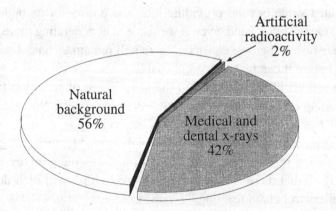

Fig. 52. Approximate proportions of the radiative exposure of typical residents in the United States. (The total 2% of the artificial sources includes the very small fraction of less than 1% of radiation escaping from nuclear power plants.)

"If you mean: 'does our understanding of radioactivity have any positive aspects?'" 'I paraphrased, "then the answer is a resounding 'yes'. Having established what the radioactive decay products and half lives of various radioactive isotopes are, we can determine the age of archaeological specimens with considerable precision by establishing the ratio in which two isotopes of the same elements are present. I have even read about the use of isotope ratios in dinosaur bones to support the proposition that they were warm-blooded creatures unlike the reptiles to which they had been linked in the past. In these and other ways, scientists have found many practical uses for radioactive elements."

"I guess I knew that." Beth pronounced. "Medical practitioners regularly use radioactive tracers to investigate the functioning of different organs and even farmers make use of radioactive elements to trace the effectiveness of plant fertilizers. Like so many other things discovered by scientists, they can be harmful as well as beneficial. If radiation can cause the destruction of healthy tissue in humans, why can't it also be used to destroy malignant tissue, as in the treatment of cancer? I guess such beneficial applications are only limited by our inventiveness."

"Well put." I was pleased to note. "One vexing aftermath of radioactivity, nevertheless, continues to bedevil us: what to do with the still radioactive products that remain after the energy has been trapped, the beneficial rays have been applied, and the studies using radioactive tracers have been completed?"

"Isn't that a political problem?" Beth asked. "Legislators may lack the scientific knowledge to come up with solutions, but who else can address this issue successfully?"

"Our state governors could and were, in fact, mandated to do so by Congress." I replied dejectedly. "As I'm sure you know, nothing but delays in decision making have ensued."

"Do you see any simple solution?" Beth tried to sound optimistic.

"Simple, yes. Acceptable to everyone, no." I replied without enthusiasm. "First we should separate the issue of treating high-level wastes that contain radioisotopes that emit very penetrating radiation and have relatively long half lives from that of disposing of low-level wastes that pose a much smaller threat to the environment."

"Isn't the Defense Department the main generator of high-level waste?" Beth asked and then answered her own question. "I believe that they simply store their massive wastes in remote military reservations in containers that they pray will not leak during any current legislator's lifetime."

"True," I responded, "although, ironically, much more attention is paid by the press to the disposal of much smaller radioactive wastes from nuclear power plants. This is also a very real concern, but relying on a technically ill-informed public to make a decision that requires technical solutions has led us to an impasse in which no action is taken and the wastes simply accumulate."

"People often overlook the fact that making no decision is, in its own way, a decision." Beth observed. "What are the possible solutions in your opinion?"

"For high-level wastes, the most reasonable approach is to make the radioactive isotopes relatively harmless by incorporating them in a synthetically formed mineral just like the ones that already contain radioactive elements in the earth's crust. Called *metamict* minerals, they have persisted for geological ages without causing any harm to our habitat. Shaped into cylindrical rods, the synthetic minerals could be dropped into the countless abandoned drill holes throughout the United States. This approach has been tried with some success in Australia years ago."

"Why haven't we tried it here?" Beth asked naively.

"Our government seems to favor giant, costly projects such as digging extensive underground storage vaults in the Southwest." I replied. "But, reluctantly, I must admit that U.S. scientists have not proved as helpful to legislators as they might have. Instead they have advanced pet solutions while criticizing those of their colleagues. For example, one group favored encapsulating the wastes in glasses that other scientists argued might leach out by ground waters."

"Sounds like they should all rally around the Australian proposal?" Beth offered. "What about the disposal of low-level wastes?"

"Low-level wastes are produced primarily by hospitals and research laboratories." I responded. "Although fairly large quantities of waste accumulate, such as contaminated containers, gloves, etc., their reduced radioactivity makes them a somewhat easier disposal problem. Still, they have to be transported to a suitable dumping site and the few western states that have accepted such wastes in the past are now refusing to do so."

"Well we can't abandon our medical treatments or the pursuit of promising research." Beth said forcefully. "What can we do?"

"First of all we have to overcome our attitudes regarding any waste disposal, namely, the 'not in my backyard' attitude." I went on more animatedly: "An ingenious idea was proposed some time ago to mix the low-level contaminated wastes in with various toxic chemical wastes that were being stored in supposedly leak-proof containers in underground sites. In this way, should the containers ever spring a leak, it could be located easily by a simple radiation detector."

"That sounds like a great idea!" Beth again sounded optimistic. "I'm sure that the threat to the populace would be no greater from the low-level radioactivity than from the toxic products in such wastes. Let's hope that our elected officials will gain the wisdom and courage to act on some of these suggestions."

# Chapter Eighteen:
# Breakfast of Corn Fritters

## *A Mess of Particles*

"You must be nearing the end of your book." Beth intoned the next morning. "What chapters have you got left still to do?"

"You're right." I replied somewhat sadly, since I really had enjoyed my book-writing project. "I am finishing the chapter on the new physics that describes the subnuclear structures that are being worked on at this very moment. After that, I plan a kind of epilogue to consider what the future of physics may hold. But that's it. It's almost all over."

"To cheer you up, I shall give you some delicious corn fritters," Beth tried sounding cheerful, "full of whole corn kernels surrounded by a yummy dough!"

"Sound good," I quipped, "especially with lots of hot maple syrup all over them."

"So what's new in physics today?" Beth asked mischievously. "I keep reading on the front page of our newspaper all about new quarks being discovered. What, exactly, is a quark?"

"To put things in perspective, let me remind you that 3,000 years ago natural philosophers argued over whether matter was continuous or composed of discrete and invisible *atomos*. By the end of the first 300 years of physics, we pretty much agreed that all matter was composed of invisible building blocks we called *atoms*. During the past century, we had to accept the fact that atoms were not indivisible nor invisible, and that they were composed of negative electrons swirling about a positive nucleus. We also had to accept the fact that the nucleus, in turn, was composed of *nucleons*. More recently we learned that the nucleons are composed of *quarks*. But what makes up these quarks?"

"That's what I asked *you!*" Beth reminded me forcefully.

"I'm afraid that I don't have a simple answer to your question," was my response. "We have good reasons for believing that quarks exist but it is the very nature of a quark that it cannot be isolated for direct observation. What we know about quarks we learn from observing the behavior of a bewildering array of subnuclear particles."

"If what you are about to tell me becomes any more bewildering than what you have already told me, I'm not sure I want to hear it." Beth responded. "What good is such knowledge, anyway?"

"Apart from the challenge that the unknown presents," I explained, "we believe that the structure of the nucleons may contain the cosmological secrets of how our universe was formed."

"Next you'll hint to me that this may be the way to discover where and how it all began. You physicists certainly don't lack immodesty!" Beth teased. "By the way, how did the connection between nuclear physics and cosmology come about?"

"Right around the start of the past century," I was glad to relate, "scientists became curious about how far the radiation from radioactive sources on earth extends into its surroundings. So they sent balloons bearing ionization detectors into the upper atmosphere. To their surprise, they detected more rather than less radiation at the higher altitudes. Since the amount did not depend on whether it was day or 'night, its source could not be just the sun. The only conclusion possible was that it originated outside our solar system and so it was named *cosmic radiation.*"

"Star Trek, here we come!" Beth burst out. "Do we know what this cosmic radiation is?"

"Yes we do." I replied. "Cosmic rays are primarily protons and alpha particles, but smaller quantities of carbon, nitrogen, iron, and other atomic nuclei are also present. Fortunately for all living things on earth, most of these charged particles are deflected by earth's magnetic field."

"Don't they make up a pair of huge belts of charged particles that circle the earth?" Beth asked. "And weren't they discovered by James van Allen from observations made by special rockets and satellites sent into space in the last forty years?"

"Quite correct. We call them the van Allen belts in his honor."

"Hasn't it also been established that many of the cosmic rays, particularly those having exceedingly high energies, originate outside of our own galaxy?" Beth inquired. "I seem to recall reading that most of what we know about the universe we learn from analysis of such radiations."

"Right again." I was glad to note. "The extremely high energies that these cosmic rays can have enabled them to generate subnuclear particles when they collided with atomic nuclei in photographic plates (Fig. 53). Many of these were hitherto unknown particles."

"Like, for example?"

"In 1928, the British mathematical physicist P. A. M. Dirac published a generalized version of the quantum theory that predicted the possible

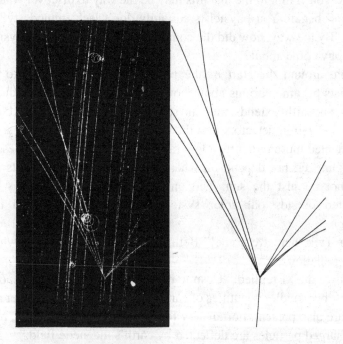

Fig. 53.   When high-energy particles pass through any medium, they leave tracks by' ionizing any atoms present (stripping off their outer electrons). The vertical lines are such tracks left by high-energy protons. Should a collision occur in which a particle is annihilated (destroyed), then other tracks, made by the nuclear particles produced in the collision, will be seen leaving the collison point, as detailed in the sketch on the right in this figure. (Photograph provided by Dr. Clifford E. Swartz.).

existence of positively charged electrons as well as the familiar negatively charged ones. Called a *positron*, such a positive electron really is an *antiparticle* of a negative electron."

"I've always been fascinated by the idea that antimatter may exist," Beth interrupted, "even if only in science fiction. What happens when an electron meets a positron?"

"Why, the two annihilate each other, of course." I replied.

"And what becomes of their respective masses?"

"The masses are converted into energy." I explained. "In this case, the energy takes the form of a high-energy photon."

"Are you pulling my leg?"

"Much as I might enjoy doing that, in this case, I am not." I went on. "As Dirac had predicted, high-energy gamma rays and other particles can produce positron — electron pairs and some of the positrons persist for some time. Sometimes called *secondary cosmic rays*, about 10% of such ionized particles on earth are positrons. At the altitudes where commercial airplanes fly, the fraction increases to 20–30% and nearly equals the fraction of negative electrons."

"Is any other kind of antimatter also created on earth?" Beth asked.

"Antiprotons were detected in 1955," I replied, "but inside particle accelerators."

"So there really is antimatter on earth!" Beth expressed genuine wonder. "Is that why we build particle accelerators? To create antimatter and study it?"

"Not exactly. The students of cosmic-ray tracks were discovering ever more strange particles, but they had to scour inordinate numbers of photographic plates to uncover them. For example, the discovery of positrons followed the examination of some 13,000 individual photographs!"

"Take, for example, the photograph that Cliff Swartz kindly sent me (Fig. 53) to illustrate nuclear collisions. The particles seen streaming vertically upward are actually antiprotons generated in a collider at Brookhaven National Laboratory. Most of the antiprotons pass right through leaving nothing but a trail of ionized atoms in their wake. (Even more zip right through without ever interacting with the atoms present.) It is necessary to examine scads of such photographs, therefore, to detect the one antiproton being annihilated by a collision with a proton (antiparticle of the antiproton)."

"What are some of the other particles making tracks in Cliff's picture?"

"Historically, some of the first particles identified were called *mesotrons,*" I replied, "because their masses fell in between those of electrons and protons and the Greek word for 'in between' is *mesos.* They were renamed *mesons* at the suggestion of a professor of classical languages who also happened to be Werner Heisenberg's father. It turned out, moreover, that there were two kinds of mesons so that they were named *mu-mesons* and *pi-mesons,* respectively."

"Aren't they also called, simply, *muons* and *pions* nowadays?" Beth asked. "You talked about muons when you described how Einstein's celebrated twin paradox was verified by comparing the half-lives of muons moving at different relativistic speeds."

"All very true." I noted proudly. "What's more, some muons have masses more than 200 times that of an electron. They can be negatively or positively charged and, like electron — positron pairs, such oppositely charged muon or pion pairs are antiparticles. Not only that, but electrically neutral pions were predicted theoretically and ultimately detected. To do this, however, particle accelerators were used."

"Sounds like the main impetus for the construction of particle accelerators may have been the preference to record controlled collisions rather than sifting through stacks of films produced by cosmic rays." Beth concluded.

"Not quite," I explained. "The need to sort through stacks of photographic or other records can't be eliminated because it takes a huge number of collisions to produce the sought-after events. The way that particle accelerators improve this process is by controlling the energies of the colliding particles. This increases the probability that a desired interaction will take place."

"So, what's the bottom line?" Beth started to fidget. "I know that lots of accelerators have been built all over the world. Some accelerate charged electrons or protons and smash them into fixed targets. Others increase the impacts by directing two particle beams travelling in opposite directions to collide with each other. But what comes out of all such possible collisions?"

"By the early 1960s, over 100 short-lived and strongly interacting particles had been identified. I stress short-lived, because most particles vanish as quickly as they are created. I stress strongly interacting because they vanish after combining with other particles or, simply, decaying into other particles. All in all, it is the trails of their interactions that tell us what has been going on."

"Wasn't it one of the leading physicists at the turn of the century, Lord Rutherford, who observed condescendingly that 'all of science is either physics or else it is stamp collecting'?" Beth now asked. "It seems to me that physicists on the trail of the quark have become stamp collectors themselves!"

## What, More Conservation Laws?

"The High-Energy-Physics community, familiarly called the HEP community, certainly was perplexed for a while." I resumed my narration. "Were these true elementary particles that were being detected or were they merely short-lived intermediary states of different types? Some theories argued that all were equally valid elementary particles. Another group claimed that some particles were more elementary than others. They got to be called 'democratic' and 'aristocratic' for supporting these divergent views."

"Meanwhile, I'm sure that the HEP experimentalists argued for building them even more powerful accelerators." Beth observed sardonically. "Surely it was very hip among physicists to favor more HEP."

"It certainly was hip to be a member of the HEP community and the brightest physics students tended to choose it as a career." I observed. "But HEP is not the only reason for constructing more energetic accelerators or *colliders* as they are now called. Particle beams have proved to be very useful in fighting malignant and cancerous cells ever since the first accelerators, called cyclotrons, were built."

"So, more colliders gave evidence of more new particles." Beth mused aloud. "How did the HEP theorists interpret all of this data?"

"The principal guide in their interpretations was the basic conservation laws of physics." I explained. "The need to conserve energy and momentum makes it possible to understand how the disappearance of one particle track must result in the creation of at least two other particles even if only one track is seen leaving the collison point."

"Here we go again," Beth interrupted. "Why must there be a second particle if the film shows no track of its passage?"

"Consider a vertical track," I went on, "like the one I showed you of a proton–antiproton annihilation (Fig. 53). At some point it stops and a new track is seen going off to the left. To conserve linear momentum in this collision, some other particle must move off to the right so that their horizontal momentum components just cancel each other. Similarly, the total energy after the collision event must be the same as it was before."

"I think I understand," said Beth. "Assuming that the visible tracks somehow disclose the energy of the particles that left those tracks, it is possible to add up the energy of the particles seen leaving a collision point and, if it is less than that of the incident particle, we know that some invisible particle carried it off. Isn't that how the existence of neutrino was first postulated?"

"Quite so." I was quick to reinforce Beth's interest. "The new quantum-mechanical theories that grew out of Dirac's elegant formulation predicted the possibility of creating an electron–positron pair from the disintegration of a single high-energy photon (Fig. 54). The electric

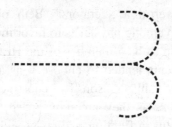

Fig. 54. A high-energy photon, like a gamma ray, can disintegrate spontaneously by giving rise to a pair of oppositely charged particles of equal mass. If this event is recorded in a photographic emulsion placed between two magnetic pole pieces, the oppositely charged particles follow identical paths that are curved in opposite directions by the interactions of their respective magnetic fields. Making use of the equivalence of mass and energy, it can also be established that they are conserved as well.

charges of these two paricles are exactly equal and opposite so that the net charge of the pair is zero. This total charge is the same, of course, as that of a chargeless photon or gamma ray that produced the pair. Thus a new conservation law is born, namely, that the sum of all the charges in the universe must remain unchanged."

"You physicists sure are quick to make up universal truths and call them fundamental laws of nature!" Beth said with a mixture of awe and envy in her voice. "How can you be so sure?"

"The law of charge conservation has proved itself invaluable in our understanding of the solid state." I responded. "Without it, many of the technological gadgets that have revolutionized our daily lives would not have been possible."

"I certainly know about these!" Beth exclaimed. "You have told me often enough about how a little chip of silicon can control the operations of my automobile, transmit my voice in a portable telephone, record any incoming messages on my FAX machine, and on, and on."

"The three laws for momentum, energy–mass, and charge conservation throughout the universe have stood physicists in such good stead, that it should come as no surprise that HEP theorists embraced them wholeheartedly." I resumed. "In fact they have gone on to invent additional conservation laws that apply only to the multitude of elementary particles that you and I shall never encounter. In a sense it could be argued that these are more like postulates than universal conservation laws. Yet we accept the three basic laws also entirely on faith. In any case, these additional conservation laws enable HEP theorists to explain how and why the various elementary particles form and even predict the existence of yet to be observed new particles!"

"Is one of these laws the conservation of parity that supposedly was broken by a pair of theoretical physicists in New York?" Beth asked. "And what, exactly, is parity?"

"Here we enter an aspect of modern physics that is abstract, abstruse, and esoteric. Like modern art or modern music, it can be seen as beautiful by those who take the trouble to try to understand it while leaving the uninitiated wondering why anyone thinks it's art."

"Having duly humbled me," Beth said mockingly, "will you now tell me, kind sir, what parity is?"

"You have heard me hold forth on symmetry in nature." I said. "When you look in a mirror, the image you see is the symmetrical reflection of yourself. A flower exhibits rotational symmetry about its axis (Fig. 55), while a honeycomb shows the symmetry of repetition by displacement, reflection, and by rotation about the six-fold axes that can be pictured to be passing through the centers of each repeated honey cell."

"And then there is the less visible symmetry that occurs in physical processes." Beth intoned. "When two billiard balls collide, they bounce off each other and conserve energy and momentum in a symmetric manner. How am I doing?"

"Very well, indeed." I smiled. "When you stand in front of a mirror, you know that your mirror image will do exactly what you do. In the same way, it is possible to show that invisible physical processes will behave symmetrically. In quantum mechanics, such symmetry is called parity."

"Example, please!"

"Very well." I responded. "When an atomic nucleus undergoes beta decay, the emitted electron is just as likely to fly off in any one direction as in the very opposite direction. This is why such equal probabilities are described by the term parity. The law of parity in physical processes, therefore, is still another example of the symmetry and beauty inherent in physics."

"So how was this lovely law broken?"

"You will recall from our last breakfast that there are two nuclear forces called the strong and weak nuclear force, respectively." I resumed. "Calculations carried out by Tsung Dao Lee and Chen Ning Yang at Columbia University showed that parity may not hold when the weak nuclear force was involved. More specifically, they concluded that parity might be violated in the beta decay of cobalt-60 nuclei. This idea was picked up by a colleague of theirs, Chien Shiung Wu, who was studying the radioactivity of various nuclei in a neighboring laboratory. Very quickly, she confirmed that the parity of beta decay was not observed by cobalt-60 nuclei. Additional proofs by others and a Nobel Prize for Lee and Yang followed."

"Once again the men got the prize." Beth sounded rueful. "I suppose that once it was shown that parity can be violated in reactions involving weak nuclear forces, other symmetry violations could not be far behind."

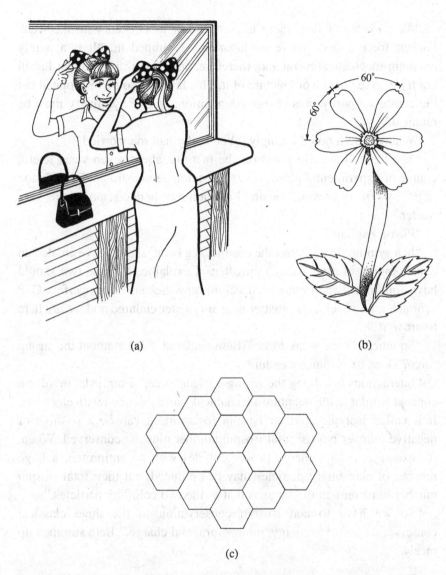

(a)

(b)

(c)

Fig. 55. (a) Reflection symmetry. (b) Rotational symmetry. Turn the flower about its stem by 60 degrees and it comes into coincidence with itself. (c) Translational symmetry. Each hexagonal cell can be repeated by translation along any direction. Note that this array also has rotational and reflection symmetries.

"As a matter of fact, there is another law associated with the weak nuclear force called charge conjugation." I jumped in. "It is a purely quantum-mechanical result and, therefore, not visualizable, so that I shall not try to give you a word picture of it. This law also could be violated but the combined parity plus charge conjugation, or C–P symmetry must be retained."

"Whew!" Beth noted jokingly. "You really had me worried."

"You laugh, but were it not for the fact that about seven years later a couple of experimental physicists at Princeton demonstrated the violation of this symmetry as well, you and I might not be here discussing this very matter."

"Please explain!"

"In a symmetric universe, the cosmic 'big bang' should have created an equal amount of matter and antimatter." I explained. "These two should have annihilated each other well before now. Because of imperfect C–P symmetry, however, more matter than antimatter endured and we are here to prove it!"

"So much for the weak force." Beth declared. " What about the strong force? Does its symmetry endure?"

"Interactions involving the strong nuclear force," I replied, "involve a concept similar to the quantum-mechanical spin associated with electrons. It is called isotypic spin, or isospin for short. It can be a positive or negative number but the total isospin number must be conserved. When, for example, a high-energy proton collides with an antiproton, a large number of elementary particles may be produced, but their total isospin number must remain the same as that of the two colliding particles."

"So we have to add isospin conservation to the three classical conservation laws for energy, momentum, and charge." Beth summed up nicely.

## Quarks and More Quarks

"The use of multiple isospin states to classify elementary particles was first suggested by Murray Gell-Mann two years after he received his Ph. D. from M.I.T. at age twenty-two. He also introduced a new kind of quantum number that he named *strangeness*, the first of several whimsical sounding terms that he has added to the HEP lexicon."

"I suppose that I should attach no more significance to this label 'strangeness' than I did to the numerical quantum numbers first proposed by Bohr and adopted by Schrödinger." Beth mused. "Is that why it's referred to as a quantum number and does it also have to be conserved?"

"As a matter of fact," I replied, "strangeness must be conserved in all interactions involving strong and electromagnetic forces but not for weak forces."

"I'm beginning to feel sorry for the weak force." Beth mockingly sympathized. "Not only is it weak, but its efforts at retaining symmetry are roundly ignored."

"By 1961, Gell-Mann had organized the elementary particles into groups of eight or ten members each." I returned to my story. "These he named the *Eightfold Way* after the teaching of the Buddha. As Mendeleyev's Periodic Table did for atomic species, gaps in these groupings suggested the existence of elementary particles yet to be observed."

"More ammunition for the HEP experimentalists to request more powerful colliders." Beth quipped. "Where is all this leading us?"

"The new particles that kept being uncovered led Gell-Mann to propose three fanciful but logical constructs." I responded. "If you accepted these three constructs, then it became possible to understand why the elementary particles fell into their respective groupings. I should hasten to add that he must have had some serious doubts about their reality because he chose to call them *quarks*."

"Isn't that a term first invented by James Joyce?" Beth popped up. "The full quote from *Finnegan's Wake* is:

> Three quarks for Muster Mark!
> Sure hasn't got much of a bark,
> And sure any he has it's all besides the mark."

"Good for you!" I marveled. "How did you come up with that so quickly?"

"You forgot that you quoted it to me just the other day from one of the books you were reading." Beth noted with a smile. "Didn't you also say that another physicist, George Zweig, had come up with a very similar construct which he named 'The Three Aces'? But how could that name compete with the quark?"

"It obviously couldn't." I went on. "The really startling aspect of these quarks that Gell-Mann and Zweig had postulated was that they had to carry fractional electric charges. These turned out to be either one-third of two-thirds and could be plus or minus."

"Didn't you tell me that the charge of an electron, *e*, was the smallest unit of charge that existed anywhere in the universe?" Beth protested. "So how can there be fractional charges?"

"Yes I did, and that is why the existence of quarks was not taken too seriously at first." I replied. "It turns out, however, that certain elementary particles, like the pions, form a group called *mesons*, and always consist of quark–antiquark pairs. Those making up the other elementary particles, called *baryons* and including protons, antiprotons, and neutrons, consist of three quarks or three antiquarks. In this way, all elementary particles still have charges expressed as multiples of that of one electron."

"This is becoming a mite confusing." Beth objected. "Can you back up and define the quarks and their roles in another way?"

"Let me try." I was eager to be understood. "The three quarks were named the *up* (*u*), the *down* (*d*), and the *strange* (*s*) quark. The electric charge assigned to them was a positive $+2/3e$ for the *u* quark, while a negative $-1/3e$ was assigned to each of the other two quarks. Their antiquarks had the same magnitude of fractional charge but with opposite signs."

"Let me see if I got this straight so far." Beth wanted to test her mettle. "A pion can consist of either a pair formed by a *u* quark and a $\bar{d}$ antiquark whose fractional charges are $+2/3e$ and $+1/3e$, respectively, or of an antiquark $\bar{u}$ and a quark *d* so that their charges add up to $-1e$. This gives us a negatively or a positively charged meson. Am I picking up this new lingo, or not?"

"You never cease to amaze me."

"Yeah! But how do I make these numbers work in sets of three for the baryons?"

"Consider a proton whose charge is $+1e$." I was quick to point out. "It is made up of three quarks *uud* whose charges are $+2/3e$, $+2/3e$, $-1/3e$, respectively, so that they add up to plus one electron."

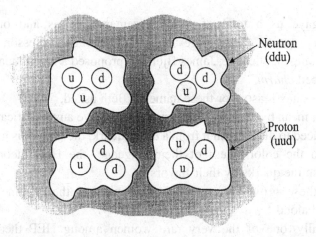

Fig. 56. Structure of a helium nucleus containing two protons and two neutrons. If we made a model of a helium atom such that the nucleus was the size of an adult human, the proton and neutron would be about the size of a hand, while a quark and an electron would be at least a thousand times smaller than that. By comparison, the entire atom would extend for some ten miles in all directions around the nucleus!

"I got it!" Beth was triumphant. "Then the antiproton must be made up of the antiquarks $\bar{u}\bar{u}\bar{d}$ whose charges all have the same magnitudes but the opposite sign so that they will add up to $-1e$. And the neutron, of course, then is made up of one $u$ and two $d$ quarks so that their combined charges add to zero! (Fig. 56)."

"I think that I have created a believer in quarks." I said smiling broadly. "In the same way, the HEP community became convinced as ever more elementary particles could be constructed out of comparable sets of quarks."

"What about an electron?" Beth wondered. "What kind of quarks are needed to make up its charge of $-1e$?"

"Electrons, muons, and neutrinos are grouped in a category called *leptons* and they are not made up of quarks or anything else as far as we know."

"Is that about it, then, for the various particles that make up the universe?"

"Not quite." I replied. "It turns out that there are two kinds of neutrinos, one kind associated with electrons or beta emissions and another with muons.

This gave us a total of four elementary leptons and only three elementary quarks. So, before long, the Harvard professor, Sheldon Glashow, and a colleague, James Bjorken, proposed a fourth quark which they dubbed *charm*."

"Is there any reason for these names?" Beth asked.

"If you mean by 'reason,' does an up quark have any significance other than the need to distinguish it from a down quark, the answer is no. In fact, to add to the colorfulness of their language, the HEP theorists now distinguish the quarks by their flavors!"

"Are these grown up men who are playing these games?" Beth wondered aloud.

"Actually, one of the very rare women among HEP theorists co-authored a review paper which suggested that, not only should it be possible to create the new $c$ quark along with its $\bar{c}$ antiquark in a collision experiment, but that scattering data already collected may have contained evidence of $c$, $\bar{c}$ pair formation. In short order, two rival groups, one at M.I.T. and another in Palo Alto, announced the discovery of a new particle that apparently was made up of this pair of charmed quarks. By an interesting coincidence, they both announced their independent discoveries on November 11, 1974. The new particle was named *psi* ($\Psi$) by one and $J$ by the other group. Well before sharing a Nobel Prize, the two discoverers agreed to name it the ($J/\Psi$) particle."

"So, now we have two pairs of leptons and two pairs of quarks." Beth recapitulated. "It seems to me that symmetry has been restored and a kind of self-satisfied feeling should have settled on the HEP community."

"Sometimes referred to as the November Revolution, what these events triggered, actually, was a new flurry of theoretical activity." I responded. "In an attempt to fit in all the known elementary particles, the emboldened theorists began assigning three *colors* to each of the quarks. This gave rise to a new theory appropriately named *chromodynamics*. One of its first predictions was that there is very little likelihood that a quark can ever be split off from its fellow quarks. It also predicted, once the various color combinations that quarks could have were considered, that there should be many more elementary particles out there waiting to be discovered."

"So, on the one hand, we are safe in painting a quark in any of three colors, since we'll never be able to isolate it, no less see it," Beth seemed to be hugely amused, "and, on the other, HEP experimentalists can clamor for ever more powerful colliders to go after all these newly invented particles! By the way, why do they need increasingly more powerful and more costly machines?"

"One of the first triumphs of this 'new' physics, as the period following the November 'revolution' has become known," I replied, "was the discovery of still one more lepton called the *tau* particle and its companion antiparticle. This lepton is almost seventeen times as massive as the muon so a collider having at least seventeen times the energy was needed to create it."

"Energy is equivalent to mass, I know." Beth was quick to interject. "This gives us an electron and an electron neutrino, a muon and a muon neutrino, and a tau. Shouldn't there also be a tau neutrino?"

"You're absolutely right. There should be. Even though it has not been discovered yet, absolutely no one admits to doubting that it's out there."

"What about the lack of symmetry now?" Beth was quick to realize. "We have six elementary leptons and their antiparticles, but only four quarks to make up the other kinds of particles."

"Very perceptive." I marveled aloud. "The discovery of the two new leptons, or at least one for sure, did launch a search for the missing two quarks. This time it took the energy capabilities of the most powerful collider at that time, the Tevatron at the Fermi National Accelerator Laboratory in Illinois, in order to find a new meson. Named the upsilon, it was quickly demonstrated that it virtually had to be made up of a new *bottom* quark and its *b* antiquark "

"And if there is a bottom, there must be a top." Beth was enjoying herself hugely.

"As day must follow night." I was caught up in her gaiety. "You may recall having read during the past year that 439 HEP scientists working at the Fermilab made a tentative announcement of their joint discovery of the top quark after carefully sifting data from proton–antiproton collisions carried out during the preceding year and a half."

"That reminds me of the off-color burlesque joke about the two hundred and forty lodge brothers who were holding an Elk's ball!" Beth laughed.

"It is certainly an incredible degree of togetherness." I agreed. "Since then, two equivalent groups finished sifting through reams and reams of

collision data before making a joint announcement that the discovery of
the top quark had been confirmed."

## The GUTS of Physics

"Let me see if I have the new physics straight," Beth requested. "There
are six leptons, including the electron, muon, and tau particle, plus their
three massless neutrinos (Fig. 57). Being elementary particles, they either
have electric charges equal to zero or that of one electron. Then there are
six quarks, carrying fractional charges, that make up all of the remaining
elementary particles. These twelve, plus their twelve antiparticles, then
are responsible for making up all the matter and antimatter found in the
universe."

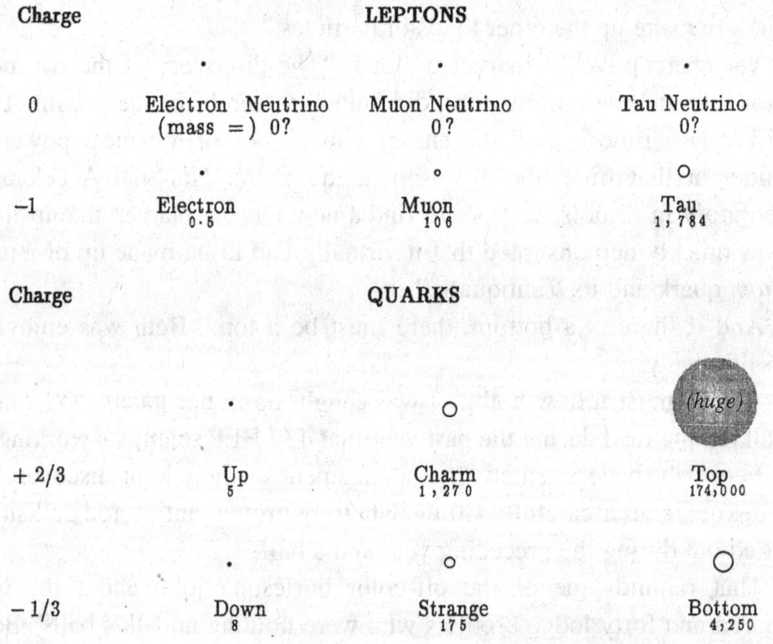

| Charge | | LEPTONS | |
|---|---|---|---|
| | • | • | • |
| 0 | Electron Neutrino (mass =) 0? | Muon Neutrino 0? | Tau Neutrino 0? |
| | • | ○ | ○ |
| −1 | Electron $0 \cdot 5$ | Muon $106$ | Tau $1,784$ |

| Charge | | QUARKS | |
|---|---|---|---|
| | • | ○ | (huge) |
| + 2/3 | Up $5$ | Charm $1,270$ | Top $174,000$ |
| | • | ○ | ○ |
| − 1/3 | Down $8$ | Strange $175$ | Bottom $4,250$ |

Fig. 57. The constituents of the Standard Model in the new physics. The names of the
constituents of matter are given above their masses, expressed as their energy equivalence
in multiples of one million electron volts.

"Excellent summary!" I felt that Beth was really beginning to appreciate the fruits of four hundred years of physics research. "This is part of the *Standard Model* of the new physics which provides the basis for our understanding of the present makeup of the universe as well as of its very origins."

"So far you have told me what the building blocks of the universe are." Beth went on. "But what holds them together and decides how they should combine?"

"Do you recall how the self-educated Faraday invented electric and magnetic fields to help him picture how electric and magnetic forces could act through space?" I asked in response. "Well, in a similar way, a Caltech professor, Richard Feynman, invented field *propagators* to show how forces could be transmitted between particles."

"Wasn't he the author of that very funny paperback called '*You must be joking Mr. Feynman*'"? Beth interrupted. "I almost died laughing at his pranks at the atomic-bomb laboratory at Los Alamos during World War II. You know, when he would figure out their combinations and open up the top secret vaults in the offices of the army brass there!"

"Feynman was a very colorful character all right. He also loved to play a pair of bongo drums and would do so with very little or no encouragement."

"But what does a field propagator *do* exactly?" Beth turned serious again.

"Although you may not have thought of it in that way," I responded, "the massless photon is a field propagator that transmits the interactions involving purely electromagnetic force fields. The signals sent out by a television transmitter are received by your TV antenna thanks to photons acting as the propagators of that electromagnetic field. Similarly, radioactive decays involving only the nuclear weak field make use of three relatively massive field propagators. Their existence can be firmly established by considering energy and momentum conservation, even if these constituents, themselves, are not readily observed."

"What makes the quarks stick together?" Beth persisted.

"Probably the most unusual concepts in all of physics." I replied. "The propagators transmitting inter-quark forces are called *gluons*, since they act to 'glue' nucleons to each other. Theirs is the most peculiar force because it becomes weaker, the closer two quarks move together, but *increases* in strength as the quarks try to move farther apart! That, of course, is why it's so difficult to pull them part. As they become separated, the energy of the gluon field increases until it converts into another quark–antiquark pair which enables the separated quarks to pair up again. The gluons are also responsible for generating the strong force that binds nucleons to each other. Moreover, these gluons also carry colors, but in color–anticolor pairs."

"This is beginning to sound like we have stepped on the other side of the mirror in *Alice in Wonderland*." Beth protested. "What are these colors and how do they relate to the one's on the quarks."

"I don't think you *really* want to know." I responded. "But, if you do, you'll have to save that question for one of my HEP colleagues. All I know is that there are eight gluons; that they can change the colors of the quarks; and, by doing all these fanciful things, theorists can explain all the possible eventualities that can arise in high-energy interactions."

"I see." Beth said with a great deal of doubt in her voice. "All these colors are really not the colors of the rainbow but just arbitrary names that HEP theorists have given to their mathematical constructs."

"That is, of course, the point exactly," I tried to sound reassuring. "The field propagators responsible for creating the electromagnetic, the weak, and the strong forces, together with the above constituent elementary particles (Fig. 57), make up the complete *Standard Model* of our current physics."

"When you consider all the propagators: eight gluons, three more for the weak force, plus the photon, and add them to the twelve elementary particles plus their antiparticles," Beth was having fun again, "you get thirty-six components of the Standard Model. Not to mention their mysterious colors. It seems to me that it's time for a simpler model!"

"You are not alone in that feeling." I observed. "Ever since Albert Einstein sought, unsuccessfully I might add, to find a theory that unified the various known forces, many theorists have continued seeking this holy grail of physics. They have had partial successes, but they have not yet been able to unify the strong, electromagnetic, weak, and gravitational forces in a single *Grand Unified Theory* that works."

"G–U–T," Beth spelled out. "It seems to me that we have reached the gut issue of physics."

"Quite so." I remarked. "There are several contenders for the title. One is called *Supersymmetry* or *Susy* for short. It is able to unify the leptons and quarks with their mediators. It even holds out the promise of quantizing the general theory of relativity which includes gravity. The only trouble is that none of its predictions has been verified so far."

"But I bet it's a truly beautiful theory!" Beth said teasingly.

"If you knew Susy, like I know Susy, ..." I couldn't resist breaking out in song. "It *is* a beautiful theory. But then there is *string theory* or *superstrings*, which is also very elegant and involves ten dimensions!"

"I can see three," Beth chimed in, "and I can accept a fourth to help understand the space–time continuum of Professor Einstein. But I don't think that I want to hear any more about a theory involving six more."

Just then the phone rang.

"That's what I call being saved by the sound of the bell at the end of the round or, at least, this breakfast!"

# Chapter Ninteen:
# Dinner at Home

## *Einstein Centennial*

"Now that you've decided to update *Physics over easy*," Beth began. "I imagine you have lots to tell me about what's been happening in the world of physics this past decade."

"A considerable change in emphasis from what physicists had pursued during most of the twentieth century." I replied. "The highlight of this period was the centennial celebration of the year 1905 when Albert Einstein published five unique papers that introduced the concept of relativity, confirmed the validity of the atomic model of matter, and helped launch the quantum revolution."

"We discussed relativity in Chapter 12." Beth reminded me, "But what were the other papers about?"

"Einstein's second paper dealt with determining atomic and molecular sizes and was accepted for his Ph.D. degree. This paper, incidentally, became the most widely cited of the five publications of that memorable year." I reported. "The third developed the theoretical model explaining Brownian motion."

"What is that?" Beth interrupted.

"A biologist named Brown had reported previously his observation of the randomly erratic motion of dust particles visible on top of a liquid placed under a microscope. What Einstein's theory described was how this motion was caused by the random movements of the molecules comprising the liquid."

"Well that sounds interesting." Beth interrupted "But why is that a big deal — particularly when compared to his paper on relativity?"

220

"Remember that, in 1905, only a limited number of scientists had accepted the idea that matter was composed of discrete atoms or molecules." I explained. "Einstein's paper demonstrated their actual existence by providing a firm proof that dust particles could be moved about by discrete molecules."

"Wasn't one of the papers about the photoelectric effect which you mentioned in Chapter 13?" Beth recalled. "And didn't Einstein receive the Nobel Prize for that paper? What's more, don't some people feel that he should have received it for the paper on relativity which is far more revolutionary?"

"He did indeed." I responded. "But how the relative importance of two such breakthroughs should be rated is beyond my ken. Special relativity introduced the concept of space–time as four interlinked dimensions within which any observable physical events are confined. It eliminated the need for an external reference frame because all motion is relative to an observer. Moreover, it is based on the postulate that the speed of light in vacuum is constant regardless of the state of relative motion of the observer. So I agree — it was revolutionary."

"But," I continued, "so was the explanation that Einstein offered for the photoelectric effect whose discoverer, Filipp von Lenard, was coincidentally awarded the Nobel Prize that very same year. Keep in mind that the quantum concept introduced by Max Planck just five years earlier had not yet taken hold in the world of physics because its only application seemed to be quite artificial and seemingly limited to producing a correct description of the emission spectra of heated metals. Even Planck harbored suspicions of its artificiality."

"What Einstein demonstrated was the duality of electromagnetic radiation in that it was a particle at the same time that it was a wave." I elaborated. "This made the quantification of energy a physical reality regardless of whether one considers the emission, transmission, or absorption of any electromagnetic radiation. Remember too that this inspired de Broglie to propose that any particle, including an electron, has an identical dual nature."

"Oh right. That was what enabled Schrödinger to formulate his wave equation and quantum mechanics was off to the races!" Beth was happy to interject. "But what was the fifth paper about?"

"That one was an elaboration on Einstein's realization that energy and mass also are interrelated, as expressed by his world famous equation $E = mc^2$. What's more," I added, "the combination of these five publications had an enormous effect on the subsequent development of physics. The emergence of quantum mechanics ultimately focused attention on atomic and subatomic particles, while the theory of general relativity helped launch the field of astrophysics, which leads to the study of the whole universe."

"Isn't that called *cosmology*? You told me it also addresses the question of how the world began." Beth went on. "It sounds like physicists split into two groups. One pursued what happens on a gigantic scale and the other on a scale too small to observe directly."

"Actually the demarcation is not so distinct." I explained. "The study of the large atomic aggregations making up the objects we encounter daily has been expanded to include unique structures consisting of just a very few atoms."

"Is that what the recently touted *nanotechnology* is all about?" Beth wondered.

"Well nanotechnology is the application of what *nanoscience* has discovered." I replied. "But before we digress further, let me conclude my description of how students of subatomic particle physics join up with astrophysicists."

"You may recall," I continued, "that experimental observation of subatomic particles requires gigantic accelerators to generate them. The difficulty of attaining the very high energies necessary to produce the more massive particles has forced physicists to look to the world far outside our Earth where processes occurring on all kinds of exotic stars involve the requisite energies."

"Does that mean that particle physicists have become stargazers?" Beth asked whimsically. "Didn't I hear about a brand new accelerator called the *Large Hadron Collider* that just went into operation in Switzerland? And wasn't it designed expressly to detect the massive Higgs particle that scientists hope will explain what mass is?"

"Indeed it was; but, on September 19, 2008, just nine days after researchers first circulated protons through the collider's 1,600 electromagnets serving to accelerate the passing particles, a melted wire produced a hole in a pipe containing liquid helium that is used to cool the superconducting electromagnets. Fifty-nine of the magnets were damaged by the spilled helium necessitating the shutdown of the collider and the postponement of the first experiment to November, 2009."

"Well it's now November of 2009 so is it up and running?"

"Regrettably, during the past year, a number of other electrical problems have been encountered so that the actual start up has been postponed again. In fact, estimates are that it will take a couple of years before the collider is fully operational so that the desired experiments can be carried out."

"Is anything being done in the meantime?" Beth asked.

"Well there is a mountain of data gathered at the *Tevatron* in Batavia, Illinois, that awaits full interpretation by the dozen or more physicists busily analyzing it. With any luck they may discover traces of the Higgs boson provided that recent estimates of its mass are on target."

"What is the heaviest particle that the Tevatron has detected so far and how does its weight compare to that of the Higgs?"

"In 1995, the Top Quark was discovered by observing its disintegration into a lighter quark and a W boson. The weight of the Top Quark, measured in units of a billion or *giga* electron volts, is 175 GeV." I replied. "This compares to the estimate of 114 to 184 GeV for the Higgs boson. So it's very likely that it may be found among the many collisions already recorded but not yet scrutinized. Completing the analysis may easily take another year. Meanwhile let's hope that the Large Hadron Collider becomes operational soon. It will, of course, take some time to sift through the date it collects."

## *A Promising Nanotechnology*

"So tell me more about this new nanoscience and the nanotechnology it has spawned" Beth requested.

"The imaginative theoretical physicist, Richard Feynman, first pointed out in 1959 that it should be possible to create novel tiny structures on the atomic scale. They could be used to manufacture tiny machines or to store data with tremendous density because their very large surface-area-to-volume ratio imparts them with unusual properties." I responded. "Twenty-seven years later, K. Eric Drexler published a book called *The Coming Era of Nanotechnology* that outlined some of the exciting possibilities that lay ahead."

"Did Feynman become the first nanoscientist?" Beth inquired.

"No, his interests lay elsewhere." I replied. "Actually various physicists and chemists took up the study and manufacture of tiny atomic aggregates like *quantum dots* that are formed atop semiconductor materials or *nanotubes* made up of carbon-atom networks rolled into tiny cylinders."

"Just how big are nanomaterials?"

"A nanometer is one billionth of a meter." I explained. "To give you a feel for that size, consider an atom whose radius is one nanometer as compared to a beach ball of one-meter radius. That is the same as the ratio of a playing-marble's radius to that of the earth."

"What on Earth can one do with such invisibly small structures?" Beth's curiosity never ceases.

"Actually the possibilities are far ranging: merely because of their reduction in size to nanometer dimensions, normally opaque copper metals become transparent while normally stable aluminum metal becomes combustible." I started to explain. "Consider next *quantum dots* which derive their name from the fact that the electrical properties of these tiny semiconductors are controlled by quantum-mechanical effects. One of their distinguishing features is their coloration. The larger the dot, the redder is its color or fluorescence. Conversely, the smaller the dot, the bluer its color becomes."

"That's all very interesting." Beth interrupted. "But of what use is that?"

"Biologists use colored dyes to help track of various processes taking place in living cells." I responded. "Because they are far brighter, quantum dots enable highly sensitive cellular imaging. They further make possible the imaging of single-cell migration which, in turn, will facilitate the observation of cancer metastasis, stem-cell therapeutics, and many other research areas."

"That sounds very promising." Beth, who relates better to biology than to physics, became more excited. "I can see where there will be many wonderful applications in medicine and in pharmaceuticals in the future. But are there any other uses for these dots?"

"There are many new industries arising to make use of their unique properties. In fact, you can buy quantum dots over the internet." I elaborated. "As to uses, a dust of such quantum dots emitting an infrared signal will adhere to any intruders helping law-enforcement agents track them down. Similarly, quantum dots can be dissolved in printing inks that will foil any counterfeiting attempts. As you see, applications are limited only by one's imagination."

"Are any other forms of nanoparticles finding practical applications and how are they created?"

"The number of different nanoparticles being manufactured for sale has risen from about two hundred a few years ago to over a thousand at present." I replied. "There are two kinds of approaches employed in their production. The top-down approach utilizes the same techniques previously used in the manufacture of semiconductor devices such as deposition from a vapor, for example. The bottom-up approach differs in that it develops novel forms of molecular self-assembly wherein atoms and molecules are coaxed into coalescing spontaneously."

"Are these exotic materials finding any application in medicine?" Beth wanted to know.

"Yes indeed." I was quick to respond. "One early sign of Alzheimer's disease is the increase of a tiny protein in the cerebrospinal fluid. By combining gold and magnetic nanoparticles that have an affinity for these proteins, it is possible to sort them out among the many other proteins present in the fluid. This screening method is about one thousand times more sensitive than conventional ones so that early detection and any possible treatment can take place at the very onset of this debilitating disease."

"But have they actually been able to apply these methods to treat disease?" Beth persisted.

"Among possible applications, one can manufacture contact lenses with nano-size pockets containing time-release drugs that cure glaucoma." I reported. "Similarly, gold-plated nanospheres have been developed that seek out tumors which typically exude more blood than healthy tissue surrounding them. An accumulation of these spheres lights up when irradiated by a low-intensity infrared light thereby locating the tumor sites. Subsequent irradiation by a more intense infrared laser beam causes the spheres to become sufficiently hot to destroy the tumor cells. I should point out that these results are derived from animal studies but field trials using human subjects are pending."

"That's really exciting." Beth exclaimed. "But is anybody considering what possible effects the release of the nanoparticles into our environment may have on living organisms?" Beth's interest in safety issues was piqued.

"As you would expect, there has been considerable interest in detecting any nanoparticles present in the environment and what possible risks they may pose. The problem is complicated by the fact that it is not the chemical composition, whose toxicity in normal-sized quantities is easily established and monitored, that is significant. Instead, the properties of the nanoparticles are determined by their size, their crystal structure, and how they interact with other kinds of particles." I observed. "To date, no protocols have been developed that encompass all these variables."

"I bet there are those who warn that any government regulations will hamper the development of this nascent industry." Beth opined. "And others who insist that the government must install safeguards to protect the populace before industrial-scale manufacturing of nanomaterials becomes a major hazard."

"You're right about the political issues. Several institutions like the Center for Responsible Nanotechnology and the Project on Emerging Nanotechnologies at the Woodrow Wilson Center are addressing and publicizing these issues."

"Meanwhile, what are nanoscientists doing to assist in these and other developments?" Beth wanted to know.

## *Quantum Entanglement*

"Probably the most promising area for future development is in solid-state quantum computation. Because of their small size, what goes on inside a dot is controlled by quantum mechanics and this makes possible such unusual effects as the *entanglement* of two or more dots."

"Hold on a minute! What is entanglement?" Beth interrupted.

"It's a translation of the German word *verschränkung* which was coined by Erwin Schrödinger in the course of explaining his *gedanken* or thought experiment involving a cat in an opaque box." I continued my discussion. "Next to the cat is a vial of poison gas which will be shattered whenever a radioactive atom, also inside the box, should decay. Because the present state of the cat is indeterminate, this state of the box can be described quantum mechanically by superimposing two wave functions: one describing a live cat and the other a dead cat."

"Doesn't that mean that the cat is both dead and alive at the same time?" Beth wondered aloud. "But how can that be? And can't we find out by opening up the box?"

"According to quantum mechanics, opening the box would cause the entangled wave functions to collapse to a single wave function describing a live or a dead cat." I replied. "As to the preceding state of superimposed wave functions, they describe a state of the cat that is indeterminate, as distinct from unknown."

"Isn't that just double talk?" Beth asked uncertainly.

"Not really." I interjected. "It's simply another manifestation of Heisenberg's indeterminacy principle which, you will recall, limits how completely we can describe the state of any event. Thus it prevents us from measuring conjugate quantities like the exact location of an electron and its momentum simultaneously. I should add that an entanglement need not be local but can take place over arbitrarily large distances."

"Here we are more than fifty years after Einstein's death still haunted by his disdain for 'spooky action at a distance' which he rejected because it required instant communication between two such distantly removed particles." Beth persisted. "And, of course, no information can travel faster than the speed of light so that instant communication is not possible."

"Actually, Einstein, along with Boris Podolsky and Nathan Rosen, proposed an experiment in 1935 that would demonstrate that the quantum-mechanical description of such action at a distance was incomplete." I observed. "Their much discussed paper has become known as the EPR Paradox. But it wasn't until 1982 that their proposed experiment was actually carried out by French physicists at the University of Paris. What they found was that Einstein and not quantum mechanics gave the wrong explanation."

"Maybe if you could give me a physical analog of an entangled pair," Beth asked in a still dubious tone, "I'd grasp this strange concept more easily?"

"Okay, here goes. As you know, the reason quantum mechanical results are considered to be weird or strange is that we don't encounter anything like them in our everyday world." I responded. "But consider a seesaw whose two opposite ends are entangled in the sense that when one is up the other must be down. Changing the position of one forces the opposite change in the other instantaneously. The quantum mechanical description of an entangled pair provides a similar linkage that causes the change of, say, the spin of one particle to respond instantaneously to any change in the spin of the other even though they may be many miles apart."

"Since they did not deny the validity of quantum mechanics, how did Einstein and his collaborators propose to explain the entanglement of two particles?" Beth wondered.

"They postulated the presence of hidden variables that are not specifically described by the existing quantum mechanics which, therefore, is not incorrect. It is just incomplete. These hidden variables must then act in some way to provide the required linkage between the possible states of entangled particles."

"Well that sounds like a reasonable possibility to me." Beth interjected. "So why did you say that the French experiment proved Einstein to be wrong?"

"Wrong may have been too strong a word." I responded. "How about unnecessary? You see, Professor Bohm suggested more recently that the two particles may be joined by a wave and they themselves are the hidden variables in the form of functions of this wave."

"That sounds good but I'm not sure it makes things any clearer for me." Beth persisted.

"I believe that your comment is based on the fact that quantum mechanics defies normal intuition which is conditioned by our experiences in a world governed by Newtonian mechanics." I elaborated. "Thus, what may have troubled Einstein is the apparent denial by quantum mechanics of what must be a physical reality even before it is established by an actual measurement. His rejoinder to the description of Schrödinger's cat by a superposition of two seemingly contradictory wavefunctions might have been that he'd like to believe that the moon was still there whether he was observing it or not."

"Well this makes me wonder: Does anyone really understand what goes on in the quantum-mechanical realm?" Beth inquired.

"When asked that same question, Professor Feynman answered that nobody really understood quantum mechanics."

"I'm not sure that's reassuring." Beth noted. "But let's return to entanglement. Tell me again what it is and how does it translate into solid-state computers?"

## Quantum Computers

"The quantum states of two or more entangled particles actually form a single entangled state which must be considered as a whole. Suppose there are two entangled particles having opposite spins symbolized by ↑ and ↓. Before determining the spin of either particle, its spin is unknown. Once it is determined to be ↑, the spin of the other particle is known to be ↓ without actually having to measure it." I elaborated. "A detailed description of how such entanglement makes computations possible is somewhat involved. But think of it in terms of the binary numbers underlying modern computer language. If the number 1 represents one state and 0 the other state, then the various combinations that can be formed by this pair become the letters, if you will, of a digital language. Analogous possibilities arise when two entangled quantum states are considered."

"Do you mean something like this?" Beth asked. "Suppose there are three bits each of which can be either a 1 or a 0. The combinations that they can form are: 000, 100, 010, 001, 110, 101, 011, and 111 for or a total of eight possibilities."

"Very good," I commended Beth. "The three bits you considered are said to form a *register*. Note that a register with 3 bits contains 8, equal to $2 \times 2 \times 2$ possible combinations. A 4-bit register contains $2 \times 2 \times 2 \times 2 = 16$ possible combinations, and so forth. What distinguishes a quantum computer is that each bit becomes a *qubit* that can store the values 0 and 1 simultaneously by quantum superposition."

"Does that mean that a register of 3 qubits can store all 8 possible combinations at once?" Beth asked eagerly.

"That's right." I noted approvingly. "So that once a register of, say, 12 qubits is formed it will contain 4,096 superposed combinations. This will enable 4,096 parallel computations to be performed instantly whereas an ordinary computer must perform them in 4,096 individual repetitions."

"That's truly mind boggling." Beth exclaimed. "Is there any limit to what a quantum computer can do?"

"Well, first of all, there are physical limitations." I explained. "A register composed of just 250 qubits formed by, say, polarizable atoms would be capable of storing more numbers than there are atoms present in our entire universe. Then there is a need to develop new computational algorithms, as well as the requisite technology so that the new form of computations can be carried out effectively. This is why the present state of development still resides in the laboratory."

"Is there anything else this new way of using quantum entanglement makes possible?" Beth asked.

"One of the hottest areas of activity is in quantum encryption." I noted. "Historically, whenever a new form of encoding information was developed, perseverance and advances in technology provided a means for breaking the codes. Quantum entanglement enables a system of information transmission that is safe from interception because any intruder into an entangled system would simultaneously disrupt the communication and inform the sender of the intrusion. Not surprisingly, several commercial versions are already on the market."

"That's actually quite scary." Beth commented. "Because both the good guys and the bad guys can use it to make their communications safe from curious ears or eyes."

"Maybe that's why our Defense Department has allocated over twenty million dollars annually to learn more about quantum encryption." I added.

"There is one more development you may find interesting." I continued. "It is possible to use entangled pairs to eliminate a particle in one of the states at point A and have it appear at another point B even when the two points are separated by an opaque wall. This process is called *quantum teleportation*. Its feasibility has been demonstrated by actual experiments. Because it requires that the original particle be destroyed before it can appear on the other side of the wall, however, it does not portend a desirable mode for transportation in the future."

"No it doesn't." Beth agreed. "So I think I'll stick to transports governed by Newton's laws of motion."

# Chapter Twenty:
# Lunch at the Beach

## *Bose-Einstein Condensates*

"What a beautiful day!" I exclaimed from our restaurant perch overlooking the Gulf of Mexico.

"Yes, we are really lucky to be able to spend our winters here in Florida." Beth agreed. "But I've been thinking about your recent efforts to explain quantum entanglement to me and have some lingering question regarding what you described for me earlier. Other than proposing the EPR Paradox, did Einstein have anything else to say on this subject?"

"As far as I'm aware, he limited his other reflections to his private correspondence with several colleagues." I replied. "He did, however, make a significant contribution by proposing the existence of a fifth state of matter based on the quantum entanglement of a large number of like atoms."

"The fifth state of matter?" Beth sounded incredulous. "I know of only three states of matter, namely, a *gas* that condenses to form a *liquid* that freezes to form a *solid* as its temperature is progressively lowered. What are the other two states?"

"If instead of cooling one heats a gas to very high temperatures, the energized gas atoms undergo many collisions that strip off some of their outer electrons. Thus the gas becomes ionized and such an ionized gas is called a *plasma* which is said to be a fourth state of matter." I explained.

"Why does it have to be a fourth state of matter instead of, simply, an ionized gas?"

"Because a plasma, which also can be formed by stripping the outer electrons from gas atoms by placing them in a very strong electric field, has unique properties quite unlike those of a normal gas." I explained.

232

"For example, plasmas can conduct electricity whereas a gas is normally an insulator. Similarly, plasmas create a magnetic field."

"Is that because the moving ions constitute an electric current that must be accompanied by a magnetic field, as Maxwell had predicted?" Beth was recalling our earlier discussions. "But what is the fifth state of matter?"

"I'll tell you in a moment. But first let me give you a bit of background. In 1924 an Indian physicist, Satyendra Bose, sent a paper to Einstein in which he presented an alternative derivation of the radiation law first derived by Max Planck in 1900." I started to explain.

"Wasn't that the discovery of what has become the quantum theory of radiation?" Beth interrupted.

"You've got it!" I resumed my explanation. "What Bose described was a statistical calculation for a gas of identical photons. But he encountered difficulty in having his manuscript published because the final result did not differ from that obtained previously by Planck. Einstein, on the other hand, realized the importance of this alternative treatment and immediately translated the English-language manuscript into German and arranged for its publication."

"That was a noble thing for Einstein to do." Beth commented. "But, as I recall, he helped a number of his colleagues gain their well-deserved recognition."

"In this instance, Einstein followed up by publishing a couple of his own papers extending Bose's theory to include identical particles having mass." I hastened to add. "In doing this he discovered that, when a sizable number of identical atoms are cooled to extremely low temperatures, they slow down sufficiently to merge into the lowest possible energy state for the entire assembly. In other words, the atoms lose their identity and merge into a kind of super atom now known as a *Bose-Einstein Condensate* or simply *BEC*."

"Aha! So that is now known as the fifth state of matter." Beth opined. "Since a BEC forms only when the atoms have been cooled to exceptionally low temperatures, how long did it take before the first BEC was actually formed?"

"Seventy-one years." I responded. "It took that long to develop the means to lower the temperature to within 100 billionths of a degree above absolute zero."

"That's 0.0000001 degrees Kelvin." Beth noted. "How on Earth is that done?"

"In the summer of 1995, a group of scientists at the Joint Institute for Laboratory Astrophysics in Boulder, Colorado, led by Carl Wieman and Eric Cornell, succeeded in forming a coherent droplet of about 2,000 rubidium atoms cooled to this temperature." I started my lengthy description. "Here is how they did it: They introduced a gas of rubidium atoms into an evacuated glass vial surrounded by six lasers whose beams intersected at the center of the vial. The frequency of the laser light was tuned to an appropriate value for absorption by the gas atoms which received tiny jolts from the colliding laser photons. By means of a magnetic field, the rubidium atoms were tuned to absorb preferentially those photons that were traveling in directions opposite to the atoms' motion. This had the dual effect of slowing down the atoms, and thus cooling them, while gently nudging them toward the center of the vial where the six laser beams intersected. In this way the rubidium gas in the vial ultimately cooled to about 40 millionths of a degree above absolute zero."

"Wow!" Beth exclaimed. "What an incredible feat for just a few laser beams to accomplish!"

"Yes, but that temperature was still a hundred times higher than what was needed to form a BEC." I resumed. "By increasing the strength of the magnetic field enveloping the atoms, they were constrained to remain at the vial's center from where only the most energetic atoms could escape. As they did, the escaping atoms removed more than their share of the energy thereby cooling the remaining atoms still further until they all achieved the lowest possible energy for their collection. This is why this final stage of the process is called evaporative cooling."

"Isn't that the way any hot liquid like coffee is cooled in its cup?" Beth queried gingerly.

"Right you are!" I happily agreed. "The hottest liquid molecules escape as steam while the remaining ones equalize their lowered temperature through repeated collisions within their container."

"Has this feat been replicated by others?" Beth wondered.

"Actually it was. Working independently, Wolfgang Ketterle at MIT formed a BEC out of sodium atoms just a few months after the original success in Boulder. All three scientists shared the Nobel Prize in physics in 2001." I responded. "Subsequently, more than a dozen laboratories around the world have successfully created BECs."

"Why all this interest?"

"One way to depict a BEC is to think of the quantum-mechanical description of any particle, namely, by a wave packet concentrated at the particle's site but that extends throughout space. As the gas atoms coalesce to form a BEC, their individual wave packets overlap forming a single wave packet representing the BEC." I began. "Thus a BEC presents a macroscopic view of the quantum world normally hidden from our direct observation."

"How big is a BEC?" Beth wondered.

"The original drop of rubidium atoms contained 2,000 atoms and existed for about 10 seconds. Subsequently much larger BEC's have been formed that have lasted up to 3 minutes." I continued. "This makes it possible for scientists to observe quantum-mechanical waves with the naked eye."

"Has anybody actually done that?" Beth was becoming visibly excited.

"More than once." I noted. "Consider what happens when two BECs are placed side by side as shown in Fig. 58. If they behaved like ordinary gases, their atoms would tend to interdiffuse. Or, if they acted like regular solids, they would fall apart. But because they are actually each a giant quantum wave packet, they interfere to produce an interference pattern of minima and maxima just like those produced by interfering light waves."

"Does this mean that each BEC acts like a giant atom exhibiting its dual nature of being a particle and a wave at the same time?" Beth wondered out loud.

"That's exactly what is happening. What we see in Fig. 58 is the interference of two matter waves obeying all the laws of quantum mechanics." I elaborated. "Whereas normally restricted to submicroscopic particles a BEC measuring about one millimeter across can be seen by our naked eyes."

"What about the Heisenberg uncertainty principle?" Beth now asked. "Doesn't it state that one can't know an atom's exact location and energy simultaneously?"

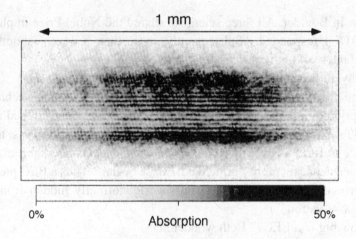

Fig. 58. Interference pattern formed by two expanding BECs. (Courtesy of Prof. W. Ketterle and *Science** magazine.)

*Andrews *et al.*, *Science* 275: 637–41 (1997).

"That's a very good question." I noted. "According to Einstein, all the atoms in a condensate must have the lowest possible energy but, according to Heisenberg, they can't then be at the exact center of the vial containing them. Furthermore, quantum mechanics posits that these atoms must have one of a set of discrete, allowed energy values wherein the lowest is not zero, just very small. At this *zero-point energy*, the atoms move slowly near but not exactly at the center of the vial. In this way, the condensate provides a rare opportunity to observe the Heisenberg principle in the macroscopic world."

"Has anyone made any use of a BEC and, if so, for what purpose?"

"The primary use for BECs has been in observing quantum mechanics in action." I replied. "A number of investigators have studied the seemingly curious features of a BEC. For example Lene Hau, a Harvard University physicist, made an interesting discovery. Assisted by two graduate students, she first prepared a BEC of sodium atoms and then passed a laser light beam through it to set up the conditions necessary for the subsequent retardation of a second laser light pulse. As she demonstrated in 1999, this second light pulse can be slowed down to a speed of just one-hundredth of a mile per second as compared to the speed of 186,282 miles per second with which light travels through the vacuum in outer space. Two years later her group actually stopped a light pulse completely for a short time."

"That's incredible." Beth exclaimed. "But how does one change the speed of light in view of Einstein's postulate that the speed of light must be constant and the same for all observers?"

"The special theory of relativity indeed specifies that the speed of light in vacuum, $c$, is constant for stationary observers, as well as, for observers moving relative to the light. But all that really means is that nothing can travel faster than $c$ — moving slower is allowed. You may recall from our discussions years ago while I wrote the original book, and as is shown in Fig. 30 of the present edition, a beam of light entering a medium at an angle to its surface is slowed and its path altered by refraction." I continued. "The speed with which the wave fronts move is called the *phase velocity* of the light and it is slowed only slightly by the refractive index of any medium. Concurrently, a macroscopic wave packet describing the light pulse moves at a *group velocity* and this can be made considerably less than the $c$."

"Is this another example of quantum weirdness?" Beth couldn't resist.

"Yes, it is a purely quantum-mechanical concept." I replied seriously. "A different trick was performed by a group including Cornell and Wieman that made a BEC shrink in size or implode. Subsequently it exploded in a way mimicking the occasional explosion of a *supernova* as observed with the aid of telescopes. The physicists therefore dubbed this explosion whimsically the *Bosenova*."

"Ah, after the *bossa nova*, the name of a musical style introduced fifty years ago in Brazil. That makes me want to dance."

# Chapter Twenty-One:
# Lunch at Venetian Bay

## *Our Energy Problems*

"What a beautiful view of the inland waterway" I remarked as we sat down for lunch at a restaurant on the shore of Venetian Bay.

"Yes indeed." Beth responded. "The bay is so clear that you can see the mullets swimming in it. But did you read in today's paper about another spill of sludge from a coal plant in Alabama?"

"I did, and it points out one of the hazards associated with using coal-fired plants to generate electricity. This spill, however, bad as it is, released only 10,000 gallons of slurry into a local stream."

"Only 10,000 gallons!" Beth said ironically. "I wonder how much harm that will cause the residents of Northeastern Alabama?"

"Potentially quite a bit, but it is considerably smaller than the one that happened three weeks earlier in Harriman, Tennessee where 5.4 million cubic yards of coal ash were spilled into local waters and countryside." I reported. "That's enough to cover about 3,000 acres with a foot-deep layer of coal ash. Moreover, this fly ash is far more toxic than the slurry released in Alabama because it contains sufficient quantities of lead and thallium to cause extensive birth defects and nervous and reproductive system disorders."

"How does this tragedy compare to the nuclear reactor accident that took place at Three Mile Island in 1979?" Beth wondered. "The panic it caused virtually stopped the construction of additional nuclear power plants in the United States."

"Despite the ensuing hoopla, the fuel meltdown caused by a failed water circulating pump and subsequent operator errors did not kill anyone and may have released enough radioactive gas to cause one case of cancer." I noted.

"By comparison, the tide of ash flowing into Alabama destroyed three houses in a nearby residential area, polluted the neighboring Emory River, and inundated roads and railway tracks throughout the large area affected."

"Well, should we abandon using coal?" Beth asked quite seriously. "Wouldn't that also greatly improve the environment and help slow global warming."

"Unfortunately, giving up coal is not realistic because it's too cheap and too plentiful." I said. "What we badly need, however, are stricter rules on controlling the emissions and the waste produced by coal plants."

"Waste products — aren't they one of the factors that militate against the use of nuclear power plants?" Beth wondered. "How would you go about disposing of their radioactive waste?"

## Energy Wastes

"You may recall that we discussed this matter previously in Chapter 17. Since then, Professor Richard Muller presented a very illuminating analysis in a book he whimsically titled *Physics for Future Presidents*."

"Well, what are some of his recommendations?"

"By considering first the contents of the radioactive wastes and the half-life of the most active components he finds that it would take about 10,000 years for their radioactivity to return to its value in the original ore." I related. "Since it's impossible to guarantee that any storage site will safely last ten millennia, people tend to throw up their hands and declare the problem intractable."

"But is giving up a realistic approach to take?" Beth asked. "No one can even begin to predict what life on Earth will be like in 10,000 years."

"Realistic, it is not, but as is the case in dealing with many technical issues, attention has been focused on the wrong question. What's more, the nuclear waste has already been produced so that not storing it is no longer an option."

"What about the proposed storage site in Yucca Mountain in Nevada?" Beth persisted. "Why are we not proceeding with its construction?"

"Because the public and its elected representatives in Washington have been making unreasonable demands for its safety." I explained. "For example, over a year ago, the U.S. Environmental Protection Agency decided to extend the regulatory period from ten thousand to one million years! But, more importantly, in 2009, the Obama administration finally decided to abandon the Yucca Mountain repository entirely after twenty-two years and $13.5 billion had been spent on researching that site."

"Was that a sensible decision or was it politically motivated?"

"The fact that the majority leader of the Senate, Senator Reid of Nevada, opposed implementation of that site may have been a factor." I replied. "But the official reason given was that the site was unusable because of its unfavorable geology."

"And discovering that took twenty-two years and over thirteen billion dollars?"

"As Professor Muller pointed out, the basic fallacy was to seek assurance of zero leakage of radioactive waste into the environment. A more realistic approach would accept a risk of 0.1% leakage rather than zero. After all, that's only one chance in a thousand. Given that the present level of radioactivity in the above-ground storage vessels is about 1,000 times higher than that of the original uranium ore, the product of these two numbers, namely 1, shows that this actual risk is about the same as it was before the uranium had been removed from the ground. Even more encouraging is the fact that the level of radioactivity in the storage containers decreases each year — dropping by a factor of 100 after 300 years."

"How does this compare to the dangers posed by the uranium that is currently present in the ground?" Beth wondered next.

"The radioactivity of the uranium contained in the mountains in Colorado, for example, will last for more than 10 billion years — not a mere few hundred — before its present level drops by a factor of ten. Moreover, much of this uranium is water soluble. As also pointed out by Muller, the ground waters passing through these formations in Colorado ultimately reach the Colorado River and become the waters being drunk throughout much of the West from San Diego up through Los Angeles."

"So, really, it appears that storing a suitably confined radioactive waste in Yucca Mountain poses less risk to the public than that presently associated with drinking a cup of tea in Los Angeles." Beth smiled.

## The Good News about Nukes

"Professor Muller makes another worthwhile suggestion." I resumed our consideration of nuclear reactor safety. "He recommends the deployment of *pebble-bed* reactors. They bury the uranium fuel inside a bed of heat-resistant pyrolitic graphite pebbles coated with silicon carbide. Such pebbles moderate the neutrons released by U-235 in the fuel and control the chain reaction thereby. Should the reactor start to overheat, for whatever reason, the more energetic neutrons that are released will be absorbed by the more plentiful U-238 atoms which, in turn, release far fewer neutrons causing the chain reaction actually to slow down!"

"That sounds too good to be true," Beth exclaimed. "Are you saying that a pebble-bed reactor is incapable of causing a meltdown like that at Three-Mile Island in 1979 or seven years later in Chernobyl?"

"Because the laws of physics govern what actually takes place in the nuclear chain reaction, a pebble-bed reactor is virtually independent of human or instrumental failures." I noted happily.

"Are there any other good news related to nuclear reactions?" Beth wondered.

"There is an interesting aftermath to the nuclear bomb tests that began in the early 1950's and went on for about ten years. Until the Limited Test Ban Treaty was signed in 1963." I responded. "One of the consequences of above-ground bomb tests is an increase in the amount of the radioactive isotope carbon-14 present in the atmosphere. Normally produced by cosmic rays striking occasional nitrogen atoms in the atmosphere, the level of C-14 nearly doubled by 1963, as illustrated schematically in Fig. 59. After that it declined in a known way as recorded by several laboratories."

"What's so interesting about that?" Beth interrupted.

"A couple of physicists at the University of Vienna using radioactive carbon dating to determine the age of archeological findings agreed to apply their technique to establish the respective dates of death of two elderly sisters who had been dead for several years when their bodies were found in 1992." I reported. "By determining the ratio of the amount of C-14 in fat cells produced in their bones just before dying to the amount

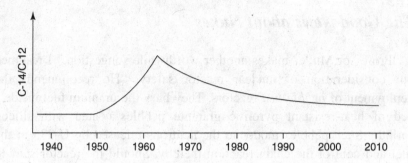

Fig. 59.   Schematic ratio of C-14 to far more abundant C-12 in the atmosphere. Note that the ratio began to rise after above-ground atomic-bomb testing increased in early 1950s and gradually declined after the Limited Test Ban Treaty was signed in 1963.

of C-12 present, it proved possible to show that the cells in one body were formed in 1988 and a year later in the other. This proved to be invaluable in adjudicating their legacies since the sisters were quite wealthy."

"That's interesting." Beth interjected. "Is there more to this story?"

"In 2002, Professor Jonas Frisén at the Karolinska Institute in Stockholm assigned the task of using this 'bomb-pulse' dating method to Kristy Spalding, a recent postdoctoral fellow from the University of Western Australia. Her task was the dating of neurons in the brains of slaughtered horses to determine whether they generated new brain sells throughout life or not."

"That sounds familiar." Beth's psychological antennae became activated. "Wasn't there a controversy between a Harvard and a Princeton professor over that very issue?"

"Yes, Pasko Rakic used to claim that there was no neurogenesis in the brain while Elizabeth Gould found a chemical tracer present in the hippocampus of terminal cancer patients. The hippocampus is the part of the brain responsible for memory and learning so finding the tracer there indicated the formation of new cells." I continued. "Once Spalding developed the very laborious technique of sorting out the neurons from horses' brains, she extended it to examining the brains of humans."

"So what did she find?"

"Based on the measured carbon ratios, Spalding and her colleagues found that the neurons in the optical cortex controlling vision were of the same age as those in the rest of the human." I replied. "They subsequently examined the human neocortex and came to the same conclusion about the absence of neurogenesis there."

"Well, what does our son, the neurobiology professor at Columbia University, have to say about this controversy?" His mother wanted to know.

"According to David Sulzer, neurogenesis definitely takes place in the olfactory bulb of the hippocampus of adult human brains but, so far, he feels there is no convincing evidence for neurogenesis in the human cortex."

"Has this atomic-pulse technique of carbon dating been applied to solving other problems?" Beth asked next.

"Spalding used it to determine that fat cells in humans are replaced every eight years. This contradicted the previously held belief that they remained the same throughout life." I responded. "Among other applications, this process also has been used to identify cell turnover in several critical organs, to authenticate the vintage of fine wines, as well as, to assist in forensic investigations."

"What effect does the increased amount of carbon entering the atmosphere and causing global warming have on the ratio of C-14 to C-12 you showed me in Fig. 59?" Beth wondered.

"The rising amount of carbon in the atmosphere from various sources on Earth dilutes the excess C-14 introduced by the earlier bomb tests so that the bomb-pulse dating method will cease to be viable."

"What does the field of physics have to tell us about global warming?" was Beth's next question.

"The physical causes of global warming are certainly governed by the laws of physics" I responded. "But they tend to take a back seat in addressing the issues when compared to the roles assumed by politicians and various entities often promoting self-serving solutions. For an interesting discussion of the science and politics of global warming I recommend Professor Muller's book."

# Chapter Twenty-Two:
# Breakfast at the Beach

## *How It All Began*

"In the Epilogue to the original edition of this book," Beth remarked while waiting for our breakfast to be served at the beach-side restaurant, "you presented a brief glimpse of what was known at the time about the origin of our universe. I suspect considerably more has been discovered since."

"Yes indeed," I responded. "The somewhat speculative ideas we had fifteen years ago have been augmented by extensive observations carried out by telescopes and other instruments mounted on high-altitude balloons and artificial satellites in outer space. We now have a much better idea of how it all began."

"As I recall from what you told me in the original edition," Beth recounted. "The entire universe began as a fiery ball about thirteen billion years ago. This was based, in part, on Hubble's observation that the universe is expanding. This, in turn, allows us to assume that the present state of the universe is like a single frame of a movie reel recording this expansion. Were we to roll the movie backwards, it would become apparent that the universe coalesces into what must have been such a fiery ball."

"Yes, this was our understanding of the *Big Bang* model of the creation of our universe" I agreed. "You also may recall that, as early as 1922, the Russian physicist Alexander Friedman had used Einstein's general relativity theory to predict three possible outcomes of such an expanding but gradually slowing universe (See Fig. 60)."

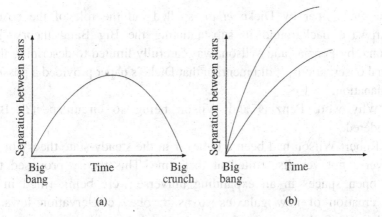

Fig. 60. Friedman's three models: (a) The universe begins with a Big Bang, expands for a while, and then slows until it collapses in a Big Crunch. (b) Two alternative models in which the expansion slows without stopping."

## Cosmic Microwave Background

"What became of the prediction by George Gamow in 1948 that a Big Bang should have left behind a kind of afterglow of microwave radiation?"

"There is a human interest side to that story." I replied. "A couple of scientists at the Bell Telephone Laboratories in New Jersey were plagued in 1965 by a mysterious background noise detected by the antenna of a radio telescope that they were using to communicate with a satellite in outer space. Unable to establish its origin, even after cleaning up the messes left behind by doves nesting in the antenna, they were forced to conclude that space was permeated by a uniform radiation in the microwave range of frequencies."

"Weren't there also scientists at nearby Princeton University, at about the same time, preparing to seek the microwave radiation that Gamow had predicted?" Beth interjected.

"After the Bell Telephone scientists Arno Penzias and Robert Wilson became aware of these attempts by Robert Dicke and his colleagues at Princeton, the two groups got together and quickly agreed to describe their respective attainments in two companion publications" I resumed.

"The first paper by Dicke *et al*. spelled out the role of the cosmic microwave background in substantiating the Big Bang theory. The second, by Penzias and Wilson, was carefully limited to describing their actual observations. It did mention that Dicke's paper provided a possible explanation."

"Why were Penzias and Wilson being so circumspect?" Beth wondered.

"Robert Wilson had been a believer in the steady-state theory of the universe not yet discredited at that time. This theory proposed that the open spaces in an expanding universe were being filled in by the creation of new galaxies so as to obey conservation laws." I explained. "So he was reluctant to appear to be a sudden supporter of the Big Bang theory. Nevertheless, despite this reluctance, Wilson gladly accepted his share of the Nobel Prize awarded to him and Penzias in 1978."

## Big Bang

"What, exactly, does the Big Bang theory predict?"

"The Big Bang theory begins about one second after the initial act of creation when the protons, neutrons, neutrinos, and other particles constituting the primordial soup had cooled sufficiently to enable the formation of atomic nuclei. It says nothing about what took place in that first second. For that reason it is sometimes referred to as the morning-after theory."

"What do you mean when you say that the primordial soup had cooled sufficiently?" Beth wondered. "How do we know what the temperature was at that time?"

"The ability to calculate what the condition was when the universe first formed from what we know about it now, enables us to estimate the temperatures and pressures that must have existed back then." I explained. "For example, when the universe was 100,000 years old, the temperature was about 6,000 °K or like that of our sun today. When the universe was only a week old, its temperature was seventeen million degrees, and one second after creation it was about ten billion degrees or what we estimate it to be at the center of a supernova explosion of a dying star."

"But let's get back to the Big Bang theory." I continued. "Its early advocate, George Gamow, and his colleagues assumed that the formation of atomic nuclei took place in the first few minutes. Although they mistakenly believed that the nuclei of all the atoms could be formed then, we now know that only the nuclei of the three lightest atoms could have been formed by this *Big Bang nucleosynthesis*. The heavier atoms were formed in stars much later."

"So hydrogen, helium, and lithium nuclei were formed in the first few minutes following the Big Bang" Beth noted. "What does that tell us about what followed?"

"The ratios of atomic nuclei formed that Gamow calculated agrees quite well with their relative abundance in our present universe. So this was a strong argument favoring the Big Bang theory." I explained. "Moreover, it predicted that the early universe was very uniform because the prevailing high temperature required its constituents to move about at high speeds randomly in all directions. This uniformity, by the way, persists throughout our present-day universe as well."

"How can you say that our present-day universe is uniform?" Beth asked. "Doesn't it consist of galaxies separated by void space?"

"On a local scale that is indeed the case, but looking at the entire universe as a whole we see an equivalent distribution of matter in any and every direction." I resumed my narrative. "Moreover, it is undergoing a homogenous expansion with each galaxy moving away from all other galaxies at a speed that is proportional to their separation. This means that the farther a galaxy is from us, the faster it moves away from us, as first noted by Hubble."

"Is this the Hubble Law I recall reading about?" Beth wondered. "But where is the center of the expanding universe?"

"The homogeneous expansion predicted by the Big Bang does not have a center or an outer edge." I explained. "The only limitation on what is the *observable* universe is imposed by the finiteness of the speed of light. By using a *light year*, which equals the distance that light travels in one year as a unit of intergalactic distance, the farthest we can see in any direction is about 13.7 billion light years. What lies beyond that we shall never be able to see because nothing can travel faster than light."

"Returning to your description of the expanding fire ball," Beth interjected, "what were the particles that it contained?"

"Initially it consisted of protons, neutrons, neutrinos, as well as other subatomic particles. It also contained energy in the form of photons which were retained within the fire ball by inward scattering from the other particles present. As the universe expanded, their speeds slowed which, in turn, caused the universe to cool. By the time it was 380,000 years old, it had cooled and expanded to such a degree that the photons, previously confined to the expanding fire ball, were able to escape it. As the universe continued to grow larger, these photons formed a uniform background that has persisted to the present day."

"I bet that's the microwave radiation background that Penzias and Wilson detected!" Beth exclaimed.

"Quite right" I noted. "By now, however, the Big Bang has been augmented by an attempt to explain what took place in the first second following the creation."

"How can we possibly tell what actually took place?" Beth wondered out loud.

"You're right to question the trustworthiness of our knowledge" I responded. "The general theory of relativity can carry us back in time a long way but at the temperatures and pressures that had to exist in the first second, relativity is no longer valid and we need to invoke the Grand Unified Theories that were first proposed about 1970."

"Do they allow us to describe what happened at the moment of creation?" Beth interrupted.

"No theory can because at that instant the temperature and pressure become infinite." I resumed my discussion. "At the exceedingly high temperatures and pressures that prevailed when the universe was barely born, even protons and neutrons, the building blocks of atomic nuclei, were unstable. Recalling that they are composed of quarks, which interact by means of gluons, we need to consult the standard model of nuclear physics to estimate what took place."

"Okay, so tell me about it."

## Inflationary Universe

"A young MIT graduate in particle physicist, Alan Guth, while in his second postdoctoral stint at Cornell University, became interested in applying the nascent Grand Unified Theory, or GUT, to determining what happened when the universe was about $10^{-39}$ seconds old. His calculations showed that the universe was expanding at an exponential rate during a tiny fraction of that first second!"

"I know that an exponential rate is one that goes faster and faster" Beth interjected. "Can you give me a more specific picture of what you mean in this case?"

"The current belief is that the universe grew from an initial size of about one hundred billionth the size of a proton to that of the order of a meter in just $10^{-35}$ seconds."

"Does this mean that the expansion rate exceeded the speed of light?" Beth asked. "Doesn't that contradict the laws of relativity?"

"Not really" I replied. "What Guth discovered was that all of space was expanding at this incredible rate during a time interval lasting from approximately $10^{-39}$ to $10^{-34}$ seconds after the birth of the universe."

"I don't understand how you differentiate between the expansion of space and the movement of the particles within that space."

"Think of an inflatable balloon whose surface is covered by evenly distributed tiny dots." I tried to come up with an easy-to-visualize model. "As the balloon expands, the dots move apart uniformly. Concurrently, a light wave connecting two neighboring dots is stretched by the same amount as the space represented by the expanding balloon. This changes the wavelength of the light wave but not its speed so that the laws of relativity are not affected by the expansion."

"Oh this is like the Doppler Effect we talked about a while back (Chapter 10)." Beth was pleased to note. "But what does this extremely rapid expansion accomplish that the original Big Bang does not?"

"There are several matters that the Big Bang could not explain." I responded. "One of them is the flatness of space that has been confirmed by a succession of satellite-borne observatories specially built to study the cosmic microwave background discovered by Penzias and Wilson.

The calculations by Guth unequivocally predict a flat universe in full agreement with these observations."

"Thinking back to a universe expanding like the surface of a balloon whose curvature progressively flattens as the balloon expands, I can see how it would appear to be flat to an observer after the balloon has expanded to cosmic dimensions." Beth commented.

"That's an astute observation." I noted with pride in Beth's grasp. "Another matter that the Big Bang cosmology couldn't explain is how distant regions at opposite ends of the visible sky could have been in causal contact with each other as demanded by their respective uniformity."

"Is that because any communication between them traveling at the speed of light would require more time than the age of the universe?"

"Exactly!" I was happy to note. "What is now known as the *Inflationary Cosmology*, proposed by Guth and augmented by others, implies that all regions of the tiny universe were in contact with each other before the first $10^{-39}$ seconds had ended and the inflationary period began, so that their like uniformity is actually expected. Inflationary Cosmology also explained why magnetic analogs of individual electric charges called *magnetic monopoles*, that were supposed to have been created by the Big Bang, had escaped detection so far. They simply had become too diluted by the end of the inflationary expansion to be observed at present despite the extensive efforts made to find them."

Largely satisfied with that explanation Beth now asked: "What force or energy was responsible for driving the inflationary expansion?"

"There is more than one possibility, so I don't have a definitive single answer." I had to admit. "First, however, let me digress to discuss the relation of energy to gravity."

"Remember that according to Newton gravity is an attractive force between any two separate bodies, that is, it is a positive force. Though this concept of an action-at-a-distance was revolutionary in the seventeenth century, it was replaced by that of a gravitational field in the twentieth. The values of the gravitational field at any point surrounding a body is the gravitational attraction that a unit of mass placed at that point would feel. Whereas the force of attraction remains positive, it can be shown that the potential energy associated with the gravitational field at that point is actually negative."

"As you can imagine, the infinitesimal speck which was expanded by inflation required the creation of a vast amount of new matter." I continued. "This, in turn, was accompanied by a vast amount of negative potential energy spreading out throughout the space that was expanding at an exponential rate. Using Einstein's relation equating mass to energy, the increasing mass or positive energy of the new matter thus was essentially balanced by the negative gravitational energy. This means that the total energy remained unchanged as required by the conservation laws of physics."

## Probing the Cosmic Microwave Background

"Does the inflationary theory have anything to say about the cosmic microwave background?" Beth now wondered.

"Actual observations of the CMB have provided considerable support for Inflationary Big Bang Cosmology." I was pleased to relate. "Early observations from high-altitude balloons confirmed the uniformity of the CMB initially reported by Penzias and Wilson. Subsequently NASA built the Cosmic Background Explorer, abbreviated, to COBE, and in January 1990 it reported that the energy density distribution in the CMB fit that of black-body radiation at a temperature of $2.735$ °K. This confirmed the predictions of the Big Bang Inflationary Cosmology to an exceptional degree of accuracy."

"Please remind me," Beth asked, "what black-body radiation is."

"A black body is one that absorbs all light incident on it." I replied. "It neither reflects nor transmits any incident electromagnetic radiation. But this fictitious black body emits light whose spectrum changes as its temperature increases from absolute zero. This emitted spectrum is called black-body radiation."

"Was there other support for the Big Bang that came out of the COBE mission?'

'The Big Bang requires there to be small temperature variations in the CMB that are related to the clumping of matter that we now observe in the sky.

Detecting these very slight variations in the CMB temperature distribution requires extremely sensitive thermal probes." I reported. "By 1992, COBE sent back a full-sky temperature map that showed the sought-after temperature variations. In 2003, the Wilkinson Microwave Anisotropy Probe produced an even higher resolution map."

"Isn't Wilkinson the scientist who likened the CMB to a fossil of the early universe?" Beth interjected.

"Yes. He observed that deducing the early universe from the CMB is like reconstructing a dinosaur from a study of its fossil bones." I replied. "But returning to the findings of the WMAP, its map enabled scientists to confirm that the universe is 13.7 billion years old and that it took 200 million years for the first stars to appear. Most importantly, WMAP established that only 4% of the universe is composed of visible matter, 23% of an invisible *dark matter*, and the remainder of a mysterious *dark energy*."

"Wow! Are you saying that the 100 billion galaxies that we can observe populating the universe account for only 4% of what's out there?" Beth exclaimed. "I find that incredible."

"In 1933, Fritz Zwicky realized that the gravitational attraction of the visible stars was insufficient to enable the formation of the galactic clusters he was observing. He postulated, therefore, that a much larger mass had to be present as well. Since it was not visible, he dubbed it *dark matter*. We are quite certain that this dark matter is not made up of the same kinds of particles that constitute visible matter. Speculation aside, we don't know its makeup but its presence throughout space has been confirmed by observing how dark matter's gravitational field deflects light passing near it as well as by other astronomical evidence."

"So you're saying that we can't see it but we can sense its presence." Beth observed. "What about this dark energy?"

"Increasing observational evidence points to the existence of a third form of matter-energy in the universe." I responded. "It is not predicted by any of our present theories although it is sometimes linked to the *ad hoc* cosmological term that Einstein invented in 1917 to counterbalance the prediction of an expanding universe by his general relativity theory. You see, Einstein and most of his colleagues believed that the universe was actually static."

"When Hubble some years later showed that the universe was not static but expanding," Beth interjected "didn't Einstein tell Gamow that introducing this cosmological term was the biggest blunder he ever made?"

"That's possibly true but, being a repulsive term, it fits the bill for a repulsive dark energy." I resumed. "Although fitting well with many observations, the calculated magnitude of the cosmological term does not."

"Well is there a better candidate?" Beth asked next.

"Particle theorists like Guth favor associating dark energy with the energy of the vacuum." I reported. "According to quantum mechanics, a vacuum is not totally empty but may contain *virtual particles*. The Heisenberg uncertainty principle makes their appearance possible but it also prevents us from ever observing them directly. Consequently, such particles may appear and disappear yet they have an energy associated with them. The calculated magnitude of this *vacuum energy*, however, turns out to be excessively large."

"Are you implying that we have no idea what dark energy is?"

"As Charles Bennett once observed, it may be more correct to say that we have too many ideas."

# Chapter Twenty-Three:
# Dinner Under the Stars

## The Stars in Heaven

"One advantage of eating out of doors is the lower background sound level that makes conversation so much easier at our age." I observed while seated in the restaurant's patio on a clear and balmy evening.

"I agree. Also watching the stars emerge in the sky adds to the sensual pleasure." Beth noted. "This reminds me of our last discussion of the cosmic microwave background. Did you say that it wasn't the only discovery supporting the Inflationary Big-Bang theory?"

"Actually, the proverbial smoking gun was provided by the study of supernovae by two competing groups of astronomers."

"That sounds intriguing." Beth declaimed as our waiter approached. "Tell me more after you order the wine."

"Before I go into that, let me review the present state of our knowledge about the life of stars in our universe." I began. "The first stars are estimated to have been formed a couple of hundred million years after the Big Bang. The space between the existing stars in a galaxy contains clouds of gas consisting mostly of hydrogen molecules which are the principal constituents of the stars as well. Local disturbances in this cloud of gas, originating from a variety of possible sources, cause the denser parts to break up into fragments which coalesce forming even denser plasmas."

"How large are these fragments?" Beth interrupted.

"Typically fragmentation stops by the time the total mass of a plasma is of the order of that of our Sun." I responded. "Then, as it continues to attract additional gas molecules, it becomes a near-spherical rotating *protostar*. The accretion of even more hydrogen molecules causes the protostar's temperature, pressure, and density, to increase until it is sufficiently high to enable the fusion of hydrogen nuclei, first into deuterium and ultimately into helium nuclei."

"Isn't hydrogen fusion the process taking place in our sun?" Beth wondered.

"Yes it is." I agreed. "In fact, once hydrogen fusion begins, the protostar becomes a full-fledged star."

"Is that based on actual observations or is it mere conjecture?"

"The actual formation of stars smaller than our Sun has been observed during the past century within our own galaxy." I responded. "The formation of much larger stars, however, is less well known."

"So, how old is the Sun and how long is its expected to last?" Beth asked next.

"Our Sun was formed about five billion years ago and it is expected to last about five billion years more."

"But how do we know that?"

"Like the existence of any star, the Sun's is based on a balance between the gravitational forces that draw the hydrogen molecules composing it inward and the pressure exerted outward by this gas of highly energized hydrogen molecules. Similarly to a toy balloon you can squeeze between your hands, the inward pressure you exert is opposed by the outward pressure of the gas contained in the balloon." I explained. "By knowing the size and density of the Sun, we can calculate the amount of hydrogen it contains. From our understanding of nuclear fusion we can next determine how long it will take to burn up all of the hydrogen present."

"So, I suspect that turns out to be five billion years." Beth concluded. "But now I'm wondering, who actually determined that hydrogen fusion lights up the sun?"

"Arthur Eddington was the first to suggest this as a possibility in 1920. Eight years later, Gamow calculated just how much energy is required to overcome the Coulomb repulsion between hydrogen nuclei so they could fuse to form deuterium. Finally, in 1939, in a Nobel-prize winning analysis, Hans Bethe examined the different possibilities for fusing hydrogen nuclei into helium"

"So what happens to the Sun after it has used up its hydrogen supply?"

## Red Giants and White Dwarfs

"After most hydrogen nuclei have fused into helium nuclei, the increased outward pressure will cause the Sun to expand to roughly a hundred times its present size. Three of the helium nuclei can fuse to form a nucleus of a carbon atom and that nucleus, in turn, can fuse with a helium nucleus to form an oxygen nucleus. Since all fusion processes release energy, the enlarged Sun keeps burning for about another billion years as a *red giant star*, so named because of its reddish color when viewed through a telescope."

"Are there many red giants in our galaxy?" Beth asked. "And what happens to them after they burn all of their helium nuclei?"

"Red giants can be found in globular clusters of a hundred thousand or more stars inside our Milky Way. They originally formed as bright Sun-like stars about ten or more billion years ago." I related. "After a red giant fuses most of the helium nuclei into carbon and oxygen, it gradually shrinks down to about the size of our Earth, as the inward gravitational pull becomes dominant. It is then called a *white dwarf*. The density of a white dwarf is about 70,000 times the density of Earth which is four times denser than the Sun. At such a density, a quantum-mechanical degeneracy of the constituent electrons creates the pressure that counters the gravitational squeeze trying to reduce the star's size further."

"I'm not sure that I understand why the red giant shrinks to the size of a white dwarf or why its density should become so huge."

"Although some of the mass of the original star converts to energy and is lost by radiation during the fusion of hydrogen and, later, of helium, the total mass decreases only slightly. In a star like our Sun, the gravitational attraction of its constituent atoms is counterbalanced by the thermally stimulated motion of the hydrogen and helium nuclei. As the red giant exhausts its supply of nuclei to fuse, gravitational attraction forces gain the upper hand and the star is reduced in size. Since its mass is reduced only slightly, its density, determined by the ratio of mass to volume, must increase."

"Wait, is the electron degeneracy in a white star like that in a Bose-Einstein condensate?" Beth interrupted.

"Yes, you can use the same term. In a BEC, electron degeneracy caused all electrons to occupy a single energy state with a blurring of our ability to determine their exact whereabouts. In a white dwarf, by comparison, it prevents a further decrease in the shrunken volume that is available to individual electrons. Keep in mind that, according to the Pauli Exclusion Principle, two electrons whose energy is described by four identical quantum numbers can not occupy the same space. Thus the degenerate electron gas exerts a quantum-mechanical counter pressure to the gravitational inward-directed pressure."

"I'm guessing that astronomers have actually seen white dwarfs in our universe," Beth wondered. "So what becomes of them ultimately?"

"Sirius, the brightest sun-like star seen in the night sky above Earth, is accompanied by a white dwarf named Sirius B. Its mass is 1.05 times that of the Sun while its size is comparable to that of Earth. That makes Sirius B about 730,000 times denser than the Earth." I recounted. "As to what becomes of white dwarfs, the answer is more speculative because of the relative youth of the universe. As fusion in the remaining gas on its periphery diminishes, an individual white dwarf will cool gradually becoming a dark inert star that will be increasingly difficult to observe. However, white dwarfs are also found to occur in pairs. Such binary dwarfs may interact explosively to produce a type Ia *supernova* about which I'll tell you later."

"Okay. But do all stars undergo the same fate as they grow older?"

"That depends on their size." I responded. "Stars who's mass is less than about eight solar masses generally cannot support fusion of atomic nuclei much heavier than helium. Therefore they will turn into red giants, becoming white dwarfs ultimately as they age."

"Are the resulting white dwarfs all alike?"

"You're not alone in wondering about that, dear. A seventeen-year old Indian physicist pondered the same question while traveling to Cambridge University in England."

"In 1928, Subrahmanyan Chandrasekhar had met Arnold Sommerfeld, a leading theoretical physicist visiting Madras, India. Sommerfeld acquainted him with the latest developments taking place in quantum mechanics at that time. In his subsequent readings Chandrasekhar encountered the publications of R. H. Fowler and Arthur Eddington, both at Cambridge University. From Eddington's book published in 1925, Chandrasekhar learned about the difficulty in reconciling the huge density of white dwarfs with the outward pressure from agitated atoms within the star as calculated using classical mechanics. But Fowler's publication in 1926 suggested that quantum-mechanical degeneracy of the electrons might provide the correct explanation instead. This idea took a while to take hold because most astronomers were unfamiliar with the just emerging quantum mechanics, a factor that was also destined to haunt Chandrasekhar for the next several years."

"What exactly did Chandrasekhar do?" Beth asked impatiently as she was finishing the main course of her seafood dinner.

"Confined for eighteen days to the boat carrying him from India to England, Chandrasekhar passed his time calculating the way that the density, pressure, and gravity change as one moves inside of a white dwarf from its surface. In other words, he worked on establishing the inner structure of a white dwarf. As Fowler had before him, he used the laws of quantum mechanics to conclude that the confined electrons were moving at speeds approaching that of light. Chandrasekhar, however, realized that this meant that he also needed to include the laws of special relativity. The mesh of the two theories was just beginning to be worked on at that time by theoreticians in Europe so it is not surprising that the nascent graduate student was unable to achieve a complete mesh. He did establish, however, that any increase in the energy of an electron already moving that fast may not cause its speed to increase further but instead served to increase the electron's inertia."

"As I remember it," Beth piped up, "inertia is the property of a mass at rest that maintains it at rest or continues to move a mass at a steady speed for ever unless it is acted upon by an external force. So how can one change its inertia?"

"Good question!" I exclaimed. "What happens when the speed approaches that of light is that the mass of the moving electron effectively increases and thereby resists any further acceleration. I should add that by the time I studied physics a quarter of a century later, that was already well known and applied in many practical devices."

"What you're saying is that Chandrasekhar was ahead of his time." Beth observed. "But is that all he accomplished?"

"His calculations led to two separate results." I resumed. "One showed how to determine the inner structure of a white dwarf. The other was his discovery that no white dwarf could have a mass larger than that of 1.4 Suns. Calculation of the inner structure was easier to understand so that it was published within a year after Chandrasekhar arrived in England. The result of meshing quantum mechanics with relativity was both more startling and more difficult to understand so it was held up by the astronomers reviewing the manuscript submitted to a British astronomical journal. So, after tiring of the extensive delays, Chandrasekhar sent the manuscript to an American astronomical journal. Fortunately it was reviewed by a physicist familiar with quantum mechanics who, after consulting an astronomer colleague, recommended its publication. Fifty-five years had to pass before Chandrasekhar was awarded the Nobel Prize for this and his subsequent discoveries."

"Well I wonder what happens to aging stars whose mass exceeds the mass of eight Suns?" Beth asked as she finished her after-dinner coffee. "But I think I'll wait to learn your answer until we get home."

# Chapter Twenty Four:
# After Dinner at Home

## *Supernovae and Neutron Stars*

"As I mentioned before, what we know about the detailed history of large stars is more speculative." I began after we settled into our twin Eames chairs and began sipping our favorite libations. "We do know that the larger the star, the briefer is its lifetime. We also know that a star whose mass exceeds about eight solar masses can support fusion processes beyond those forming carbon and oxygen."

"Speculate on and tell me more."

"Stars with masses exceeding eight solar masses can achieve temperatures high enough to trigger the fusion of carbon and oxygen nuclei causing them to form nuclei of heavier elements. For example, the outermost layer of a large star has a temperature of about five million degrees like that in our Sun. This is sufficient to ignite the fusion of hydrogen into helium." I began my explanation. "Moving inside the star, the temperature and pressure rises, until it is high enough to ignite the fusion of helium into carbon and oxygen. Deeper still, successive layers form containing oxygen, nitrogen, magnesium, and so forth, until reaching an inner core made up of radioactive nickel-56 which subsequently decays into stable iron-56. That's as far as spontaneous nuclear fusion can go."

"Why does the fusion process stop with iron?" Beth naturally wanted to know.

"Because the binding energy per nucleon in iron is higher than that in any other element, the fusion of iron-56, just like its fission into lighter elements, requires additional input of energy from an external source." I explained.

"Hold on a second so I can get it straight. Were you suggesting earlier that a heavy star has a structure resembling an onion except that concentric spherical layers are composed of ever heavier atoms?" Beth now asked.

"That is what astrophysicists believe the structure of actual heavy stars to be," I agreed, "with the fusion processes occurring within the boundaries between the layers. The innermost nickel-iron core resists the inward gravitational pressure of the outlying layers by means of outward-directed degeneracy pressure of the electrons within it. As the core accumulates ever more iron, its size will exceed the Chandrasekhar limit causing it to implode and form a *neutron star* while the outer layers explode in what is classified a Type II supernova."

"Wow! That's exciting!" Beth exclaimed. "I suppose that supernovae have been observed by astronomers for many years."

"The first recorded supernova was reported by Chinese astronomers near the end of the second century." I resumed. "As you can imagine, the sudden appearance of a new bright star that lasts for a few weeks and then fades away has been attracting attention ever since. Thus Fritz Zwicky, an irascible physicist at Caltech, was drawn by supernovae to the study of astrophysics in the early 1930s. After Chadwick's discovery of the neutron in 1932, Zwicky proposed that the core of an aging star might be caused to implode until it reached a density akin to that in an atomic nucleus. The matter in such an ultra dense core would transform into a compressed neutron gas whose gravitational force would shrink its size and mass. The 'lost' mass would then power the explosion of the supernova."

"What was the physical basis for Zwicky's proposal of a neutron star?"

"Imaginative intuition and crude calculations that ultimately turned out to be quite prescient." I responded. "Joined by Walter Baade at the nearby Mount Wilson Observatory, Zwicky continued observing supernovae and advancing speculative ideas about neutron stars. Zwicky's conjectures largely ignored by his contemporaries at that time; were vindicated about thirty-five years later when neutron stars were actually identified."

"I smell another interesting tale in the offing."

"The year is 1967 and the place is Cambridge University. A graduate student named Jocelyn Bell Burnell is examining reams of chart paper generated by a radio telescope designed to detect radio signals emanating from outer space. She notes a regularly recurring signal appearing about once every second. In consultation with her advisor, she identifies its source as a rapidly rotating neutron star and submits her discovery as the core of her doctoral dissertation. When the paper describing that finding was published, her thesis advisor's name appeared first among five co-authors, as is often the case in such publications. That's how Antony Hennish got to share the 1974 Nobel Prize — but not with Jocelyn Bell Burnell."

"I imagine that caused quite a furor." Beth commented. "I find it difficult to believe that such chauvinism still existed in the Swedish Academy as recently as that."

"Jocelyn Bell Burnell received many other awards and distinctions, however, culminating with the award of the Order of the British Empire in 1999 and her elevation to Dame Commander in 2007."

"Can you describe what a rotating neutron star is like?"

"A neutron star is an extremely dense sphere, about 10 km in diameter, which spins rapidly about an axis. In this case its rate of spinning was about one revolution per second although it can be orders of magnitude larger. As a result, it is surrounded by a very strong magnetic field whose magnetic axis is inclined to its axis of rotation (Fig. 61) for the same

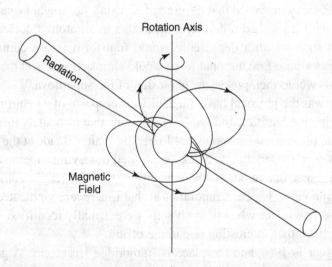

Fig. 61.   Schematic representation of a spinning neutron star, its magnetic field, and beam of radiation emanating along the magnetic axis.

reason that the Earth's magnetic axis and rotational axis are inclined to each other." I began my description. "A beam of electromagnetic radiation streams out along the magnetic axis as it rotates so that the neutron star appears like a rotating light beacon atop a lighthouse. Should this rotating stream strike Earth, it will do so in recordable pulses so that it is alternately called a *pulsar*. So far, we have observed radio, visible-light, and x-ray pulsars."

"Do all massive stars become neutron stars in their old age?" Beth asked whimsically.

## Black Holes

"Stars who's mass exceeds about twenty solar masses form extremely dense inner cores. When these cores exceed about 2.5 solar masses, no known repulsive force can push back with sufficient force to overcome the gravitational pressure of the surrounding layers. The core implodes therefore and forms a *black hole*."

"You mentioned black holes when discussing Einstein's general relativity theory." Beth interjected. "But what is the current concept of black holes and how did it evolve?"

"What follows is based on theoretical calculations rather than on direct observation." I began my answer. "The imploding stellar core is believed to compact to the size of a point having infinite density. Called a *singularity*, its immense gravitational force keeps even light photons from escaping once they have crossed inside the *event horizon* marking the circumference at which the velocity necessary to escape from the gravitational pull, called the *escape velocity*, just equals the speed of light. This is the reason, of course, why it is not possible to observe a black hole directly.

The mathematical singularity first appeared in calculations based on Einstein's general relativity theory carried out by Karl Schwatzschild while he was serving in the German army during WWI. He sent two papers to Einstein in 1915. The first described the curvature of spacetime near and inside an isolated star. The second paper, completed shortly before he died of an illness contracted on the Russian front, described the formation of what is now called the *Schwarzschild singularity*. Although Einstein found the concept of a black hole bizarre, he introduced the papers to physicists on behalf of their author."

"I'm assuming we no longer consider this concept so bizarre." Beth opined.

"In 1939, J. Robert Oppenheimer and Hartland Snyder, using Einstein's general relativity, concluded that massive stars ultimately became black holes. John A. Wheeler, their eminent colleague at Princeton, however, considered these calculations unconvincing until 1962, by which time he also agreed that some massive stars must collapse into black holes."

"Is there no way to actually detect any black holes that might exist in our universe?" Beth wondered aloud.

"It was left to a Russian astrophysicist, Yakov Borisovich Zel'dovich, whose theoretical calculations had contributed significantly to the evolution of our present understanding of black holes, to initiate an actual search for them in 1964." I responded. "This was based on looking for the ways in which black holes could be expected to affect any stars in their vicinity as well as the interstellar gases outside the event horizon."

"I would think that a black hole's gravitational pull would have a pronounced effect on its surrounding matter." Beth volunteered.

"In the decade beginning in 1965, calculations by a new wave of young astrophysicists showed that black holes can spin and create a tornado-like swirling motion in the spacetime surrounding them. A star or other massive object falling into a black hole would cause it to pulsate in the same way that the Earth pulsates during an earthquake." I continued. "Their calculations also demonstrated that a black hole's mass, electric charge and rate of spin were the dominant parameters in determining the shape of its event horizon and its gravitational pull. By 1975, however, they concluded most of their analyses and the members of this group gradually drifted off to study other problems. About that time, the sixty-five year old Chandrasekhar decided to expand and refine the complex mathematics the group had developed. Eight years later he published the result in an epic tome titled *The Mathematical Theory of Black Holes*."

"Have the Hubble space telescope and other satellite-based observatories helped in the hunt for black holes?" Beth now asked.

"Indeed they have." I responded. "We now know that giant black holes lie at the centers of most galaxies including our own Milky Way. We've also detected black holes in more distant galaxies."

## Supernovae as Standard Candles

"Earlier you promised me you would describe Type Ia supernovae." Beth reminded me. "How do they differ from Type II supernovae?"

"Type II supernovae are produced by massive stars when their extremely dense cores implode and form a neutron star." I explained. "The supernova is made up from the exploding outer layers containing all the elements formed out of the hydrogen in the outermost layer. By comparison, the Type Ia supernova is produced by an exploding white dwarf which, as you may recall, is the residue of a red giant star which has fused all of its hydrogen and helium ultimately to form carbon and oxygen. Thus their spectra exhibit no hydrogen emission lines."

"Am I correct in stating that the reason that the fusion process stops at oxygen is because the precursors of white dwarfs are stars whose mass is no larger than several solar masses?" Beth asked. "But what causes a white dwarf to explode?"

"It is believed that isolated white dwarfs like Sirius B and, ultimately, our Sun will not explode. Instead they'll become progressively dimmer." I began. "When a neighboring white dwarf or gas can supply additional mass to a white dwarf, it grows larger, hotter and denser until it reaches the Chandrasekhar limit. At that point it explodes in a thermonuclear explosion. I should point out that no one has actually observed this process evolving so that the preceding is based entirely on theoretical calculations."

"I assume that the name supernova is based on the extreme brightness of the exploding star." Beth observed. "How can we tell which type it is when we observe a supernova?"

"Supernovae as bright as four billion Suns have been observed. But the brightness we measure on Earth depends strongly on how far we are from the exploding star." I responded. "Determining the supernova type is made possible by recording its emission spectrum. If any hydrogen lines are observed we know it can't be Type Ia supernova since white dwarfs have fused virtually all of their hydrogen long before their explosion. By comparison, Type II spectra are much richer and include some of the strong emission lines from hydrogen."

"How *do* we measure the distance to a supernova or to any star in a distant galaxy for that matter?" Beth picked up on my comment about a star's brightness.

"That's a question that has intrigued astronomers for many years. Different stars were considered, but a reliable marker for galaxies outside the Milky Way was not identified until recent times when the origins of supernovae became better understood. Because white dwarfs are believed to have similar histories, their explosions are expected to produce comparable brightness. Thus they can serve as standard candles whose red shift then provides a means for determining the distance from an observer on Earth." I began. "Two groups of astronomers set out to verify this: The *Berkeley Supernova Cosmology Project* led by Saul Perlmutter on the west coast, and, on the east coast, the *High-Z Supernova Search Team* led by Robert P. Kirshner from Harvard. The Berkeley group concentrated on maximizing the number of Type Ia supernovae recorded in order to minimize statistical variations while the High-Z Team focused on optimizing the accuracy of their smaller number of observations."

"What is the nature of the possible differences between two observations of the same supernova?" Beth asked for clarification.

"Several factors can complicate the analysis. Chief among them is the presence of interstellar dust that partly absorbs the transmitted light beam." I resumed. "Whereas the Berkeley group had observed almost three times as many distant supernovae, their error of measuring each was almost twice as large as that in the High-Z Team's data. Interestingly, although the two groups worked independently and even competed with each other for primacy, their final results were remarkably similar."

"Was there a final winner in this competition?" Beth wondered.

"Because the final results of their measurement were somewhat startling, both groups were reluctant to publicize them before further verification." I reported. "Ultimately, in 1998, the High-Z Team decided to submit its paper to *The Astronomical Journal*. After that, the Berkeley group submitted its similar results to *The Astrophysical Journal* but the paper did not appear until 1999."

"What was so startling about their results?' Beth naturally inquired.

"Both teams were skittish because their measurements indicated that the distant galaxies were not moving away from us at a speed that increased with their remoteness, as everyone had believed, but their departures were actually accelerating!" I explained. "The only way this can be explained is to postulate that there must be a dark energy driving the galaxies apart. This new unknown joins the invisible dark matter whose existence, as you already know, has been confirmed independently by measurements of the Cosmic Microwave Background."

"So is this the end of your story?"

"I prefer to think it's the beginning of the next phase of our continuing exploration of the universe. To this end, the European Space Agency launched a rocket on May 14, 2009 carrying two observatories to their respective orbits. The Herschel Space Observatory is equipped to detect far infrared and submillimeter radiation that should reveal the formation of stars hitherto unobserved because cosmic dust blocks radiation at other wavelengths. The Planck satellite will scan the cosmic microwave background with more sensitive low-temperature detectors over a much wider angular range than that reached by the Wilkinson Probe. It is also hoped to detect swirls in the CMB polarization which would confirm the correctness of the inflationary model of the creation of our universe."

"Is there a better model?" Beth now wondered.

"Several alternatives to inflationary cosmology have been proposed. They tend to make use of multi-dimensional string theory and, typically, include an initially contracting phase before passing through a singularity to an expanding phase."

"But do any of them fit observations better than inflation?" Beth interjected.

"So far, none of the alternatives proposed are either as well developed as inflationary cosmology nor do they fit observations as well. But additional observational evidence should prove helpful. It will take the Plank probe nearly two years to collect and process an excess of 100 billion measurements — but the quest continues."

# Glossary of Physics Terms

*Absolute temperature scale* — A temperature scale whose zero point equals −273° centigrade and represents the lowest temperature that is attainable. (Also called the *Kelvin* scale.) (p. 36)

*Acceleration* — Rate at which the speed of an object increases (or decreases) with time. (p. 10)

*Action and reaction* — Newton's third law of motion states that for each action (force) there must be an equal and oppositely directed reaction (counterforce). (p. 20)

*Alpha particle* — Helium nucleus emitted by a larger atomic nucleus undergoing transmutation. (p. 182)

*Ampere* — Unit of electric current equal to one electric charge flowing past a point per second. (p. 61)

*Anion* — A negatively charged atom or *ion* that has gained one or more outer electrons. (p. 166)

*Antiparticle* — A particle having the same mass but an opposite charge as that of a 'regular' particle. (p. 205)

*Atmospheric pressure* — The pressure exerted at the earth's surface by the weight of all the air above it. (p. 173)


*Band model of solids* — Graphic representation of the allowed and forbidden energy values for electrons in a solid. (p. 69)

*Barometer* — Any device that can measure the atmospheric pressure. (p. 174)

*Baryons* — Subnuclear particles of relatively large mass such as neutrons, protons, etc., consisting of three *quarks* (or *antiquarks*). (p. 214)

*Beta particle* — Electron (or positron) emitted by a nucleus undergoing transmutation (radioactive decay). (p. 182)

268

*Big Bang theory* — Cosmological theory postulating the creation of the universe in an explosive transformation of energy into mass that has been expanding thereafter. (p. 130)

*Black hole* — A massive star that has collapsed (shrunk) due to its gravitational force which is sufficiently strong to prevent any visible light from leaving the very dense mass. (pp. 130 & 236)

*Bohr model* — Planetary model devised for hydrogen in which the central positive nucleus (proton) is circled by one electron that can travel in fixed or *stationary* orbits corresponding to the electron's allowed energies. (p. 143)

*Bose-Einstein condensate (BEC)* — State of matter consisting of atoms, confined by laser beams at temperatures close to absolute zero, and forced to occupy the lowest available quantum state for the assembly. (p. 234)

*Brownian motion* — Random motion of any particles suspended in a liquid because they are buffeted about by the atoms or molecules making up the liquid. (p. 158)

*Buoyant force* — Upward force exerted by a fluid on any object immersed in the fluid and equal to the weight of the displaced fluid. (p. 174)

*Calory* — Amount of thermal energy required to raise the temperature of 1 gram of water 1° centigrade. (p. 34)

*Cation* — A positively charged atom or *ion* that has lost one or more outer electrons. (p. 166)

*Centigrade scale* — Temperature scale whose zero point is set at the temperature of melting ice and 100° centigrade corresponds to the boiling point of water. (p. 35)

*Centripetal force* — Force directed toward the center of a circular path followed by the object on which the force acts. (p. 22)

*Chain reaction* — A self-sustaining reaction that is maintained by the release of more reaction activators (neutrons in a uranium pile) than minimally needed to continue the reaction process. (p. 195)

*Chaos theory* — Theory based on nonlinear equations whose solutions change in an unpredictable way after a known change in the starting conditions of some event.

*Charge conservation* — Principle requiring that the electric charge within a system remains constant even while the system may be changing in some other way. (p. 209)

*Chromodynamics* — Theoretical extension of *quantum mechanics* to subnuclear particles that assigns colors as a kind of *quantum number.* (p. 216)

*Colliders* — Particle accelerators in which particles moving in opposite directions are caused to collide with each other. (p. 207)

*Compton effect* — The different frequencies of x rays scattered at two different angles by light elements like carbon are explained by assigning a particle-like behavior to x rays scattered at one angle and wave-like diffraction to those appearing at the other angle. (p. 154)

*Conservation laws* — Fundamental laws of nature based on postulates that the quantity being conserved (total mass, energy, etc.) remains constant (unchanged) throughout the universe. (p. 15)

*Corpuscular theory of light* — Newton's theory postulating that light consisted of particles or corpuscles. (p. 105)

*Correspondence principle* — A new theory must agree with the results of any verified theory that it replaces or supplants. (p. 128)

*Cosmic Microwave Background* (*CMB*) — Afterglow of microwave radiation from the Big Bang. (p. 245)

*Cosmic rays* — Particles bombarding the Earth's atmosphere from sources in outer space. (p. 203)

*Covalent bond* — The sharing of pairs of electrons by two adjacent atoms to form an exceptionally strong interatomic bond. Also called an *electron–pair bond.* (p. 168)

*Defects* — Any departures from ideal atomic arrays in crystals, such as inserted or missing atoms, for example. (p. 228)

*Diffraction* — The bending of a traveling wave around an obstacle in its path. (p. 98)

*Doppler effect* — The compaction (or rarefraction) of sound waves produced by a source moving relative to an observer. (p. 102)

*Dynamo* — Precursor of modern-day electric generator consisting of wire loops mechanically rotated within a magnetic field. (p. 84)

*Efficiency* — Ratio between output and total input, usually expressed as a percentage. (p. 38)

*Electric current* — Flow of electrons through some medium expressed in *amperes*. (p. 76)

*Electric potential* — A measure of the electric potential energy imparted to a charge. Also called a voltage because it is measured in *volts*. (p. 60)

*Electric power* — Rate at which electric energy is consumed in doing work and expressed in *watts* or *kilowatts*. (p. 71)

*Electric transformer* — Motionless device in which electric power is transferred from one coil to another by electromagnetic induction. (p. 83)

*Electrolyte* — Nonmetallic medium in which electric current is carried by charged atoms *(ions)* instead of electrons. (p. 60)

*Electromagnetic field* — Intertwined electric and magnetic field moving with the speed of light (in vacuum). (p. 88)

*Electromagnetic spectrum* — Total range of frequencies of electromagnetic radiation starting with very small values of radio waves and going on to microwaves (radar), infrared, visible light, ultraviolet, x rays, and gamma rays with increasing frequency. (p. 89)

*Electron* — Negative particle carrying the smallest possible electric charge. (p. 52)

*Electron gas* — Name given to the collection of nearly free electrons permeating a metal and responsible for its characteristic metallic properties such as conductivity, opaqueness, luster, etc. (p. 64)

*Electric (electrostatic) attraction* — The attraction force between positive and negative charges measured in coulombs. (p. 52)

*Energy* — A measure of the ability to do work. It cannot be demonstrated except by example. (Like pornography?) (p. 30)

*Entanglement* — Quantum entanglement of two physical entities is a purely quantum-mechanical effect in which what happens to one entity determines what happens to the other. (p. 229)

*Entropy* — A measure of the disorder that accompanies the expenditure of energy. (p. 47)

*Ether* — An invisible, odorless, all pervasive medium, having no mass, that was invented in order to explain how electromagnetic waves could propagate through space. (p. 95)

*Fahrenheit scale* — Temperature scale in which the melting point of ice is +32°F and the boiling point of water is 212°F. (p. 35) another. (p. 219)

*Field propagator* — Transmitter of force from one subnuclear particle to another (p. 219)

*Gamma particle* — High-energy photon emitted by an atomic nucleus during its transmutation (radioactive decay). (p. 183)

*Geodesic* — The shortest distance on a curved surface connecting any two points. (p. 130)

*Gluon* — *Field propagator* holding *quarks* to each other with a force that grows rapidly in magnitude as they are pulled further apart so that a *quark* can never be isolated. (p. 220)

*Gravitational attraction* — Attractive force exerted by one mass upon another. (p. 24)

*GUT* — Acronym for Grand Unified Theory supposed to unify all known forces. (p. 220)

*Heat* — Energy flow from a warmer to a colder substance. (See also *thermal energy*.) (p. 31)

*HEP* — Acronym for High Energy Physics. (p. 207)

*Inert gas* — The last element in any row of the Periodic Table has a completely full outer electron shell hence it neither seeks to gain, lose, or share any of its electrons. Inert gas atoms, therefore, do not form bonds with any other atoms except under very unusual circumstances. (p. 164)

*Inertia* — Tendency of a body to resist any change in its state of motion (or rest). (p. 11)

*Inflationary universe* — Exponential expansion of universe about $10^{-36}$ sec. after the Big Bang. (p. 249)

*Interference* — Two or more traveling waves passing the same point interfere with each other *constructively*, if their amplitudes add to produce a larger displacement; or *destructively*, if the resulting displacement is decreased. (p. 100)

*Inverse-square law* — The magnitude of an interaction between two objects is inversely related to the square of the distance separating them. (p. 24)

*Ion* — Electrically charged atom. (p. 61, 166)

*Ionic bond* — Nondirectional bond formed by *anions* and *cations* in a continuous fashion inside ionic crystals. (p. 167)

*Isotopes* — Identical atoms whose nuclei contain different numbers of neutrons. (p. 184)

*Kinetic energy* — Energy associated with any motion. (p. 30)

*Latent heat* — Thermal energy consumed in changing the state of an object without changing its temperature. (p. 35)

*Leptons* — Subnuclear particles *not* made up of *quarks,* such as *electrons, muons, and neutrinos.* (p. 215)

*Magnet* — An object that has been magnetized. (p. 76)

*Magnetic field* — Region of space containing lines of magnetic force surrounding any magnet. (p. 80)

*Magnetism* — Property exhibited by a limited number of materials that can attract or repel other similar magnetic materials. (p. 80)

*Maxwell's equations* — Four mathematical expressions expressing the relationship between the electric and magnetic fields composing an electromagnetic field. (p. 87)

*Mesons* — Particles having masses intermediate between those of an electron and a proton and consisting of *quark–antiquark* pairs. May carry positive, negative, or no electric charge. (p. 206)

*Momentum* — A quantity of motion equal to the product of mass and velocity (speed). (p. 16)

*Muon* — One kind of *meson.* (p. 15, 206)

*Nanometer* — One billionth of a meter. (p. 225)

*Nanotechnology* — Application of nanoscience to development of practical devices a few nanometers in size. (p. 225)

*Neutron* — Electrically neutral *nucleon* having a very slightly larger mass than a *proton.* (p. 186)

*Neutron star* — Rapidly spinning core of a collapsed massive star, formed during a supernova, consisting of an ultradense gas of neutrons. (p. 261)

*Neutrino* — Nearly massless, electrically neutral particle first postulated to account for energy losses accompanying beta decay. (p. 189)

*Nuclear fission* — The splitting of the nucleus of a heavy atom into two or more parts accompanied by emission of radiation and release of energy. (p. 194)

*Nuclear fusion* — Combination of two light nuclei to form a heavier nucleus with an accompanying release of energy. (p. 197)

*Nucleon* — Name given to any particle making up an atomic nucleus, notably, a proton or a neutron. (p. 186)

*Paradox* — A true-sounding statement that seems to defy our common sense or usual beliefs. (p. 126)

*Parity* — Principle stating that nuclear reactions are *symmetric*. (p. 210)

*Pauli exclusion principle* — Proscribes any two electrons in a single atom having the same four quantum numbers. (p. 162)

*Periodic Table* — Mendeleyev's grouping of the elements into periods according to their atomic number Z, now known to equal the atom's total number of electrons. (p. 160)

*Photon* — Massless particle of energy (quantum) composing electromagnetic radiation and imparting it a particle-like nature. (p. 135)

*Pion* — One kind of *meson*. (p. 206)

*Polymer* — A giant molecule composed of upwards of tens of thousands of atoms. (p. 177)

*Positron* — A positive particle having the electric charge +e. It is the *antiparticle* of an electron. (p. 205)

*Potential difference* — Difference in the electric potential (voltage) between any two points. (p. 60)

*Potential energy* — Energy stored in an object that can be released to do work (cause motion). (p. 30)

*Primary colors* — Blue, green, and red. (p. 120)

*Proton* — Nucleus of a hydrogen atom having a positive charge equal in magnitude the negative charge of one electron. (p. 143)

*Quanta* — Smallest units of some quantity, notably, energy, whose smallest unit is denoted by Planck's constant *h*. (p. 135)

*Quantum mechanics* — Probabilistic description of the behavior of matter at the atomic and subatomic levels in terms of their wave functions. (p. 149)

*Quantum numbers* — Sets of interrelated numbers arising from the solution of quantum-mechanical equations that can be used to determine the allowed energies that atomic electrons can have. (p. 143, 161)

*Quantum theory* — A theory based on the postulate that the elemental unit of energy, called a *quantum,* has a fixed magnitude $h$ so that radiant energy must be an integral multiple of this quantity. (p. 135)

*Quarks* — Elementary particles believed to constitute subnuclear particles. (p. 202, 213)

*Radioactivity* — The process of emitting radiation from atomic nuclei undergoing transmutation to another kind of nucleus. (p. 181)

*Refraction* — Bending of light beam on passing from one medium to another caused by a change in its speed. (p. 109)

*Resistance* — Opposition to the flow of an electric current measured in *ohms.* (p. 61)

*Röntgen* — Discoverer of x rays. (p. 90)

*Solid* — Any substance that retains' its own shape unaided

*Sonic boom* — Crest of sound waves accompanying any moving object exceeding the speed of sound produces a shock wave which causes a sharp sound to be heard. (p. 103)

*Special relativity* — Einstein's framework for describing any physical event consists of four dimensions; length, width, height, and time. The theory of relativity includes a basic postulate that the speed of light, $c$, in vacuum is the same for all observers regardless of their state of motion. (p. 122)

*Spectrum* — Complete range of radiation of a particular kind. The *electromagnetic spectrum* encompasses all radiation from low-frequency radar to the highest energy gamma rays whereas the *visible spectrum* ranges from red to violet light. (p. 115)

*Standard model* — Total array of elemental particles constituting all matter in the universe. (p. 219)

*Strong nuclear force* — Strongest force known that binds the nucleons in an atomic nucleus to each other. It drops off very quickly with distance so that it has near-zero magnitude outside the nucleus. (p. 186)

*Supernova* — Bright light emitted by an exploding star. Type Ia supernova is formed by exploding white dwarf. Type II supernova is formed by exploding neutron star. (p. 265)

*Superposition Principle* — Because traveling waves displace *but do not carry along* the medium through which they travel, the displacements produced by two (or more) waves can be added to each other at each point along their paths. (p. 100)

*Supersonic* — Sound whose frequency exceeds the audible range of human hearing. (p. 103)

*Superstrings* — One of the two leading *GUT* models. (p. 221)

*Supersymmetry* — One of the two leading *GUT* models. (p. 221)

*Symmetry* — Equivalance between identical points on identical objects linked to each other in space by some operation repeating the one from the other. (p. 227)

*Temperature* — Measure of the average kinetic energy of the atoms (or molecules) comprising a body usually expressed in *degrees* centigrade, Fahrenheit, or Kelvin. (p. 34)

*Thermodynamics* — The science of thermal energy transfer usually expressed by two laws. (p. 39) *First law:* The total energy of any system must remain constant. (p. 43) *Second law:* Heat must flow from a hot to a cold object. (p. 45)

*Transmutation* — Transforming an element into another by changing the number of protons in the atomic nucleus. (p. 188)

*Thermal energy* — Total heat content of a body. (p. 33)

*Thermodynamics* — The science of thermal energy transfer usually expressed by two laws. (p. 39) *First law:* The total energy of any system must remain constant. (p. 43) *Second law:* Heat must flow from a hot to a cold object. (p. 45)

*Transmutation* — Transforming an element into another by changing the number of protons in the atomic nucleus. (p. 188)

*Uncertainty Principle* — Heisenberg showed that there is an inherent indeterminacy or limit in the accuracy with which it is possible to measure any physical quantity. Considered in complimentary pairs, the product of the errors in measuring two such quantities like speed and location cannot be less than Planck's constant $h$. (p. 153)

*van Allen belt* — Charged particles circling the earth in at least two separate belts maintained by the earth's magnetic field. (p. 204)

*Visible spectrum* — The eye can see colors ranging from red through violet. (p. 114)

*Wave* — An undulation that repeats itself in a regular (periodic) manner and is described by its *amplitude* (maximum displacement) and *frequency* (number of repetitions per second) or *wavelength* (distance between successive crests). (p. 97)

*Wave front* — Leading edge of a traveling wave. (p. 98)

*Wave function* — Name given to the solution of quantum-mechanical equations that describe an electron. (p. 149)

*Wave–particle duality* — First postulated by Einstein to explain the photoelectric effect, the idea that electromagnetic radiation is composed of photons having the simultaneous properties of a particle *and* a wave was expanded by de Broglie to include matter particles like electrons as well. (p. 147)

*Wavelength* — Distance between successive crests or troughs of a wave. Its magnitude is the reciprocal of its *frequency*. (p. 97)

*Weak nuclear force* — A force that is a trillion times weaker than electromagnetic forces and plays a role only in radioactive decay. (p. 187)

*Weight* — Force of gravitational attraction exerted by the Earth on any mass. (p. 17)

*Work* — Product of force exerted times the distance through which it acts. (p. 30)

*x rays* — High-frequency portion of electromagnetic spectrum. (p. 90)

# Acknowledgments

Copies of the original draft of this manuscript were read by Janis Allen, Mike McCarthy, and Dorrine and Stanley Stolar. Their many helpful comments and suggestions played an invaluable role in shaping this book and rendering it more comprehensible to the non-physicist reader. I owe them a great debt for this labor of love.

The preparation of this edition required me to learn new ways to use my computer. I want to thank Andrew and Deirdre Knapp for their invaluable aid in helping me acquire these critical skills.

The Stolars also kindly read and critiqued the additional chapters in the second edition. For this too I am deeply grateful.

My wife Beth not only fed me breakfasts while listening to my tales, but read and edited the final manuscript as well. This I acknowledge publicly although I have private ways of expressing my thanks to her.

The words that you actually read and the responsibility for what messages they bear remain mine alone.

# Index